Mock California Supplemental Exam

(CSE of Architect Registration Exam)

CSE Overview, Exam Prep Tips,
General Section and Project Scenario Section,
Questions, Solutions and Explanations

Gang Chen

ArchiteG®, Inc.
Irvine, California

Mock California Supplemental Exam (CSE of Architect Registration Exam): CSE Overview, Exam Prep Tips, General Section and Project Scenario Section, Questions, Solutions and Explanations

ArchiteG®, Inc.
http://www.ArchiteG.com

ISBN: 978-1-61265-015-9

PRINTED IN THE UNITED STATES OF AMERICA

Dedication

To my parents, Zhuixian and Yugen,
my wife, Xiaojie, and my daughters,
Alice, Angela, Amy, and Athena.

Disclaimer

Mock California Supplemental Exam (CSE of Architect Registration Exam) provides general information about California Supplemental Exam. The book is sold with the understanding that neither the publisher nor the authors are providing legal, accounting, or other professional services. If legal, accounting, or other professional services are required, seek the assistance of a competent professional firm.

The purpose of this publication is not to reprint the content of all other available texts on the subject. You are urged to read other materials, and tailor them to fit your needs.

Great effort has been taken to make this resource as complete and accurate as possible. However, nobody is perfect and there may be typographical errors or other mistakes present. You should use this book as a general guide and not as the ultimate source on this subject. If you find any potential errors, please send an e-mail to:
info@ArchiteG.com

Mock California Supplemental Exam (CSE of Architect Registration Exam) is intended to provide general, entertaining, informative, educational, and enlightening content. Neither the publisher nor the author shall be liable to anyone or any entity for any loss or damages, or alleged loss or damages, caused directly or indirectly by the content of this book.

If you do not wish to be bound by the above, you may return this book to the publisher for a full refund.

Legal Notice

ARE Mock Exam series by ArchiteG, Inc.

Time and effort is the most valuable asset of a candidate. How to cherish and effectively use your limited time and effort is the key of passing any exam. That is why we publish the ARE Mock Exam series to help you to study and pass the ARE exams in the shortest time possible. We have done the hard work so that you can save time and money. We do not want to make you work harder than you have to.

Do not force yourself to memorize a lot of numbers. Read through the numbers a few times, and you should have a very good impression of them.

You need to make the judgment call: If you miss a few numbers, you can still pass the exam, but if you spend too much time drilling these numbers, you may miss out on the big pictures and fail the exam.

There are no official CSE sample questions and answers. The existing CSE practice questions or exams by others are either way too easy or over-killed. They do NOT closely match the real CSE exam.

We have completed comprehensive research on the official CSE Test Plan by investigating many related websites, reference materials, and other available CSE exam prep materials. We formulated our mock exam questions as close as possible to the CSE Test Plan and the real CSE exam. Some readers have failed the CSE exam two or three times before they eventually passed the exam with our help.

This book includes detailed answers and explanations to questions for both the Project Scenario Section and General Section of the exam.

There is also some extra information on the ARE and exam-taking tips in Chapter One. This is based on NCARB and other valuable sources. The strategies discussed in this chapter are valuable for both the ARE and CSE. This is a bonus feature we included because we want our readers to be able to buy our ARE mock exam books together or individually. We want you to find all the necessary ARE and CSE exam information and resources in one place through our books.

How to Use This Book

We suggest you read *Mock California Supplemental Exam (CSE of Architect Registration Exam)* at least three times:

Read once and cover Chapter One and Two, the Appendixes, the related FREE PDF files, and other resources. Highlight the information you are not familiar with.

Read twice focusing on the highlighted information to memorize. You can repeat this process as many times as you want until you master the content of the book.

After reviewing these materials, you can take the mock exam, and then check your answers against the answers and explanations in the back, including explanations for the questions you answer correctly. You may have answered some questions correctly for the wrong reason. Highlight the information you are not familiar with.

Like the real exam, the mock exam includes the Project Scenario Section and General Section.

The actual CSE exam has *approximately* 100 questions. You may have a few extra beta questions in the real exam that do not count towards your score, but CAB does not divulge which questions will not be counted. We included 120 questions, 42 questions for the Project Scenario Section, and 78 questions for the General Section. There are also 20 extra questions so that you can prepare better for the real CSE exam.

Make sure you time yourself properly, especially for the Project Scenario Section since the exam time is only one hour and fifteen minutes.

Review your highlighted information, and take the mock exam again. Try to answer 100% of the questions correctly this time. Repeat the process until you can answer all the questions correctly.

CSE is one of the most difficult exams because it requires critical thinking under real project situations. After reading this book, you will greatly improve your ability to deal with the real CSE questions, and have a great chance of passing the exam on the first try.

Take the mock exam at least two weeks before the real exam. You should definitely NOT wait until the night before the real exam to take the mock exam. If you do not do well, you will go into panic mode and NOT have enough time to review your weaknesses.

Read for the final time the night before the real exam. Review ONLY the information you highlighted, especially the questions you did not answer correctly when you took the mock exam for the first time.

This book is very light so you can easily carry it around. These features will allow you to review the CSE content whenever you have a few minutes.

The Table of Contents is very detailed so you can locate information quickly. If you are on a tight schedule you can forgo reading the book linearly and jump around to the sections you need.

Table of Contents

Chapter One

Overview of the Architect Registration Exam (ARE)

A. First Thing First: Go to the Website of your Architect Registration Board and Read all the Requirements of Obtaining an Architect License in your Jurisdiction
See the following link:
http://www.ncarb.org/Getting-an-Initial-License/Registration-Board-Requirements.aspx

B. Download and Review the Latest ARE Documents at the NCARB Website

1. Important links to the FREE and official NCARB documents
The current version of the Architect Registration Exam includes seven divisions:

- Programming, Planning & Practice
- Site Planning & Design
- Building Design & Construction Systems
- Schematic Design
- Structural Systems
- Building Systems
- Construction Documents and Services

***Note:** Starting July 2010, the 2007 AIA Documents apply to all ARE Exams.*

Six ARE divisions have a multiple-choice section and a graphic vignette section. The Schematic Design division has NO multiple-choice section, but two graphic vignette sections.

For the vignette section, you need to complete the following graphic vignette(s) based on the ARE division you are taking:

Programming, Planning & Practice
Site Zoning

Site Planning & Design
Site Grading
Site Design

Building Design & Construction Systems
Accessibility/Ramp
Stair Design
Roof Plan

Schematic Design
Interior Layout
Building Layout

Structural Systems
Structural Layout

Building Systems
Mechanical & Electrical Plan

Construction Documents & Services
Building Section

There is a tremendous amount of valuable information covering every step of becoming an architect available free of charge at the NCARB website:
http://www.ncarb.org/

For example, you can find the education guide regarding professional architectural degree programs accredited by the National Architectural Accrediting Board (NAAB), NCARB's Intern Development Program (IDP) guides, initial license, certification and reciprocity, continuing education, etc. These documents explain how you can qualify to take the Architect Registration Exam.

I find the official ARE Guidelines, exam guide, and practice program for each of the ARE divisions extremely valuable. See the following link:
http://www.ncarb.org/ARE/Preparing-for-the-ARE.aspx

You should definitely start by studying the official exam guide and practice program for the ARE division you are taking.

2. **A detailed list and brief description of the FREE PDF files that you can download from NCARB**
   The following is a detailed list of the FREE PDF files that you can download from NCARB. They are listed in order based on their importance.

   - **ARE Guidelines** includes extremely valuable information on the ARE overview, six steps to complete ARE, multiple-choice section, graphic vignette section, exam format, scheduling, sample exam computer screens, links to other FREE NCARB PDF files, practice software for graphic vignettes, etc. You need to read this at least twice.

- **NCARB Education Guidelines** (Skimming through this should be adequate)

- **Intern Development Program Guidelines** contains important information on IDP overview, IDP steps, IDP reporting, IDP basics, work settings, training requirements, supplementary education (core), supplementary education (elective), core competences, next steps, and appendices. Most of NCARB's 54-member boards have adopted the IDP as a prerequisite for initial architect licensure. This is why you should be familiar with it. IDP costs $350 for the first three years, and then $75 annually. The fees are subject to change, and you need to check the NCARB website for the latest information. Your IDP experience should be reported to NCARB at least every six months and logged within two months of completing each reporting period (the **Six-Month Rule**). You need to read this document at least twice. It has a lot of valuable information.

- **The IDP Supervisor Guidelines** (Skimming through this should be adequate. You should also forward a copy of this PDF file to your IDP supervisor.)

- **Handbook for Interns and Architects** (Skimming through this should be adequate.)

- **Official exam guide, references index, and practice program (NCARB software) for each ARE division**
  This includes specific information for each ARE division. (Just focus on the documents related to the ARE divisions you are currently taking and read them at least twice. Make sure you install the practice program and become very familiar with it. The real exam is VERY similar to the practice program.)

  a. **Programming, Planning & Practice (PPP)**: Official exam guide and practice program for the PPP division
  b. **Site Planning & Design (SPD)**: Official exam guide and practice program (computer software) for the SPD division
  c. **Building Design & Construction Systems (BDCS)**: Official exam guide and practice program for the BDCS division
  d. **Schematic Design (SD)**: Official exam guide and practice program for the SD division
  e. **Structural Systems (SS)**: Official exam guide, references index, and practice program for the SS division
  f. **Building Systems (BS)**: Official exam guide and practice program for the BS division
  g. **Construction Documents and Services (CDS)**: Official exam guide and practice program for the CDS division

- **The Burning Question: Why Do We Need ARE Anyway?** (Skimming through this should be adequate.)

- **Defining Your Moral Compass** (Skimming through this should be adequate.)

- **Rules of Conduct** is available as a FREE PDF file at:
  http://www.ncarb.org/
  (Skimming through this should be adequate.)

C. The Intern Development Program (IDP)

1. What is IDP?

IDP is a comprehensive training program jointly developed by the National Council of Architectural Registration Boards (NCARB) and the American Institute of Architects (AIA) to ensure that interns obtain the necessary skills and knowledge to practice architecture independently.

2. Who qualifies as an intern?

Per NCARB, if an individual meets one of the following criteria, s/he qualifies as an intern:

a. Graduates from NAAB-accredited programs
b. Architecture students who acquire acceptable training prior to graduation
c. Other qualified individuals identified by a registration board

D. Overview of the Architect Registration Exam (ARE)

1. How to qualify for the ARE?

A candidate needs to qualify for the ARE via one of NCARB's member registration boards, or one of the Canadian provincial architectural associations.

Check with your Board of Architecture for specific requirements.

For example, in California, a candidate must provide verification of a minimum of five years of education and/or architectural work experience to qualify for the ARE.

Candidates can satisfy the five-year requirement in a variety of ways:

- Provide verification of a professional degree in architecture through a program that is accredited by NAAB or CACB.

  OR
- Provide verification of at least five years of educational equivalents.

  OR
- Provide proof of work experience under the direct supervision of a licensed architect

2. **How to qualify for an architect license?**
Again, each jurisdiction has its own requirements. An individual typically needs a combination of about eight years of education and experience, as well as passing scores on the ARE exams. See the following link:
http://www.ncarb.org/Reg-Board-Requirements

For example, the requirements to become a licensed architect in California are:
- Eight years of post-secondary education and/or work experience as evaluated by the Board (including at least one year of work experience under the direct supervision of an architect licensed in a U.S. jurisdiction or two years of work experience under the direct supervision of an architect registered in a Canadian province)
- Completion of the Intern Development Program (IDP)
- Successful completion of the Architect Registration Examination (ARE)
- Successful completion of the California Supplemental Examination (CSE)

California does NOT require an accredited degree in architecture for examination and licensure. However, many other states do.

3. **What is the purpose of ARE?**
The purpose of ARE is NOT to test a candidate's competency on every aspect of architectural practice. Its purpose is to test a candidate's competency on providing professional services to protect the health, safety, and welfare of the public. It tests candidates on the fundamental knowledge of pre-design, site design, building design, building systems, and construction documents and services.

The ARE tests a candidate's competency as a "specialist" on architectural subjects. It also tests her abilities as a "generalist" to coordinate other consultants' works.

You can download the exam content and references for each of the ARE divisions at the following link:
http://www.ncarb.org/are/40/StudyAids.html

4. **What is NCARB's rolling clock?**
   a. Starting on January 1, 2006, a candidate MUST pass ALL ARE sections within five years. A passing score for an ARE division is only valid for five years, and a candidate has to retake this division if she has NOT passed all divisions within the five year period.

   b. Starting on January 1, 2011, a candidate who is authorized to take ARE exams MUST take at least one division of the ARE exams within five years of the authorization. Otherwise, the candidate MUST apply for the authorization to take ARE exams from an NCARB member board again.

These rules were created by the **NCARB's rolling clock** resolution and passed by NCARB council during the 2004 NCARB Annual Meeting.

5. **How to register for an ARE exam?**
See the instructions in the new ARE guideline at the following link:
http://www.ncarb.org/en/ARE/~/media/Files/PDF/Guidelines/ARE_Guidelines.pdf

6. **How early do I need to arrive at the test center?**
Be at the test center at least 30 minutes BEFORE your scheduled test time, OR you may lose your exam fee.

7. **Exam format & time**
All ARE divisions are administered and graded by computer. Their detailed exam format and time allowances are as follows:

1) **Programming, Planning & Practice (PPP)**

Introduction Time:	15 minutes	
MC Testing Time:	**2 hours**	**85 items**
Scheduled Break:	15 minutes	
Introduction Time:	15 minutes	
Graphic Testing Time:	**1 hour**	**Site Zoning (1 vignette)**
Exit Questionnaire:	15 minutes	
Total Time	**4 hours**	

2) **Site Planning & Design (SPD)**

Introduction Time:	15 minutes	
MC Testing Time:	**1.5 hours**	**65 items**
Scheduled Break:	15 minutes	
Introduction Time:	15 minutes	
2 Graphic Vignettes:	**2 hours**	**Site Grading, Site Design**
Exit Questionnaire:	15 minutes	
Total Time	**4.5 hours**	

3) **Building Design & Construction Systems (BDCS)**

Introduction Time:	15 minutes	
MC Testing Time:	**1.75 hours**	**85 items**
Scheduled Break:	15 minutes	
Introduction Time:	15 minutes	
3 Graphic Vignettes:	**2.75 hours**	**Accessibility/Ramp, Stair Design, Roof Plan**
Exit Questionnaire:	15 minutes	
Total Time	**5.5 hours**	

4) **Schematic Design (SD)**

Introduction Time:	15 minutes	
Graphic Testing Time:	**1 hour**	**Interior Layout (1 vignette)**
Scheduled Break:	15 minutes	
Introduction Time:	15 minutes	
Graphic Testing Time:	**4 hours**	**Building Layout (1 vignette)**
Exit Questionnaire:	15 minutes	
Total Time	**6 hours**	

5) **Structural Systems (SS)**

Introduction Time:	15 minutes	
MC Testing Time:	**3.5 hours**	**125 items**
Scheduled Break:	15 minutes	
Introduction Time:	15 minutes	
Graphic Testing Time:	**1 hour**	**Structural Layout (1 vignette)**
Exit Questionnaire:	15 minutes	
Total Time	**5.5 hours**	

6) **Building Systems (BS)**

Introduction Time:	15 minutes	
MC Testing Time:	**2 hours**	**95 items**
Scheduled Break:	15 minutes	
Introduction Time:	15 minutes	
Graphic Testing Time:	**1 hour**	**Mechanical & Electrical Plan (1 vignette)**
Exit Questionnaire:	15 minutes	
Total Time	**4 hours**	

7) **Construction Documents and Services (CDS)**

Introduction Time:	15 minutes	
MC Testing Time:	**2 hours**	**100 items**
Scheduled Break:	15 minutes	
Introduction Time:	15 minutes	
Graphic Testing Time:	**1 hour**	**Building Section (1 vignette)**
Exit Questionnaire:	15 minutes	
Total Time	**4 hours**	

8. How are ARE scores reported?

All ARE scores are reported as Pass or Fail. ARE scores are typically posted within 5 to 10 days since 2013. See the instructions in the new ARE guideline at the following link: http://www.ncarb.org/en/ARE/~/media/Files/PDF/Guidelines/ARE_Guidelines.pdf

9. Is there a fixed percentage of candidates who pass the ARE exams?
No, there is NOT a fixed percentage of passing or failing. If you meet the minimum competency required to practice as an architect, you pass. The passing scores are the same for all Boards of Architecture.

10. When can I retake a failed ARE division?
You can only take the same ARE division once within a 6-month period.

11. How much time do I need to prepare for each ARE division?
Every person is different, but on average you need about 40 to 80 hours to prepare for each ARE division. You need to set a realistic study schedule and stick with it. Make sure you allow time for personal and recreational commitments. If you are working full time, my suggestion is that you allow no less than 2 weeks but NOT more than 2 months to prepare for each ARE division. You should NOT drag out the exam prep process too long and risk losing your momentum.

12. Which ARE division should I take first?
This is a matter of personal preference, and you should make the final decision.

Some people like to start with the easier divisions and pass them first. This way, they build more confidence as they study and pass each division.

Other people like to start with the more difficult divisions so that if they fail, they can keep busy studying and taking the other divisions while the clock is ticking. Before they know it, six months has passed and they can reschedule if need be.

Programming, Planning & Practice (PPP) and Building Design & Construction Systems (BDCS) divisions often include some content from the Construction Documents and Service (CDS) division. It may be a good idea to start with CDS and then schedule the exams for PPP and BDCS soon after.

13. ARE exam prep and test-taking tips
You can start with Construction Documents and Services (CDS) and Structural Systems (SS) first because both divisions give a limited scope, and you may want to study building regulations and architectural history (especially famous architects and buildings that set the trends at critical turning points) before you take other divisions.

Complete mock exams and practice questions and vignettes, including those provided by NCARB's practice program and this book, to hone your skills.

Form study groups and learn the exam experience of other ARE candidates. The forum at our website is a helpful resource. See the following link:
http://GreenExamEducation.com/

Take the ARE exams as soon as you become eligible, since you probably still remember portions of what you learned in architectural school, especially structural and architectural history. Do not make excuses for yourself and put off the exams.

The following test-taking tips may help you:

- Pace yourself properly. You should spend about one minute for each Multiple-Choice (MC) question, except for the SS division questions which you can spend about one and a half minutes on.
- Read the questions carefully and pay attention to words like *best, could, not, always, never, seldom, may, false, except,* etc.
- For questions that you are not sure of, eliminate the obvious wrong answer and then make an educated guess. Please note that if you do NOT answer the question, you automatically lose the point. If you guess, you at least have a chance to get it right.
- If you have no idea what the correct answer is and cannot eliminate any obvious wrong answers, then do not waste too much time on the question and just guess. Try to use the same guess answer for all of the questions you have no idea about. For example, if you choose "d" as the guess answer, then you should be consistent and use "d" whenever you have no clue. This way, you are likely have a better chance at guessing more answers correctly.
- Mark the difficult questions, answer them, and come back to review them AFTER you finish all MC questions. If you are still not sure, go with your first choice. Your first choice is often the best choice.
- You really need to spend time practicing to become VERY familiar with NCARB's graphic software and know every command well. This is because the ARE graphic vignette is a timed test, and you do NOT have time to think about how to use the software during the test. If you do not know how, you will NOT be able to finish your solution to the vignette on time.
- The ARE exams test a candidate's competency to provide professional services protecting the <u>health, safety, and welfare</u> of the public. Do NOT waste time on aesthetic or other design elements not required by the program.

ARE exams are difficult, but if you study hard and prepare well, combined with your experience, IDP training, and/or college education, you should be able to pass all divisions and eventually be able to call yourself an architect.

14. Strategies for passing ARE exams on the first try

Passing ARE exams on the first try, like everything else, needs not only hard work, but also great strategy.

- **Find out how much you already know and what you should study**

  You goal is NOT to read all the study materials. Your goal is to pass the exam. Flip through the study materials. If you already know the information, skip these parts.

  Complete the NCARB sample vignette(s) and MC questions for the ARE exam you are preparing for NOW without ANY studying. See what percentage you get right. If you get 60% right, you should be able to pass the real exam without any studying. If you get 50% right, then you just need 10% more to pass.

This "truth-finding" exam or exercise will also help you to find out what your weakness areas are, and what to focus on.

Look at the same questions again at the end of your exam prep, and check the differences.

***Note:** We suggest you study the vignettes in the official NCARB Study Guide first, and then study the official NCARB Multiple Choice (MC) sample questions, and then other study materials, and then come back to NCARB vignette and the NCARB (MC) questions again several days before the real ARE exam.*

- **Cherish and effectively use your limited time and effort**

  Let me paraphrase a story.
  One time someone had a chance to talk with Napoleon. He said:
  "You are such a great leader and have won so many battles, that you can use one of your soldiers to defeat ten enemy soldiers."

  Napoleon responded:
  "That may be true, but I always try to create opportunities where ten of my soldiers fight one enemy soldier. That is why I have won so many battles."

  Whether this story is true is irrelevant. The important thing that you need to know is **how to concentrate your limited time and effort to achieve your goal. Do NOT spread yourself too thin**. This is a principle many great leaders know and use and is why great leaders can use ordinary people to achieve extraordinary goals.

  Time and effort is the most valuable asset of a candidate. How to cherish and effectively use your limited time and effort is the key to passing any exam.

  If you study very hard and read many books, you are probably wasting your time. You are much better off picking one or two good books, covering the major framework of your exams, and then doing two sets of mock exams to find your weaknesses. You WILL pass if you follow this advice. You may still have minor weakness, but you will have covered your major bases.

- **Do NOT stretch your exam prep process too long**
  If you do this, it will hurt instead of help you. You may forget the information by the time you take the exam.

  Spend 40 to 80 hours for each division (a maximum of two months for the most difficult exams if you really need more time) should be enough. Once you decide on taking an exam, put in 100% of your effort and read the RIGHT materials. Allocate your time and effort on the most important materials, and you will pass.

- **Nailing the vignette**
  This is the easiest way to improve your chance of passing any ARE exam on the first try, and can be done in a very short period. You really need to spend time practicing to become VERY familiar with NCARB's graphic software. You need to practice it many times until you know it like the back of your hand.

  If you fail any ARE exam because of the vignette(s), you ONLY have yourself to blame.

  Note: *Again, we suggest you study the vignettes in the official NCARB Study Guide first, and then study the official NCARB Multiple Choice (MC) sample questions, and then other study materials, and then come back to NCARB vignette and the NCARB (MC) questions again several days before the real ARE exam.*

- **Resist the temptation to read too many books and limit your time and effort to read only a few selected books or a few sections of each book in detail**
  Having all the books but not reading them, or digesting ALL the information in them will not help you. It is like someone having a garage full of foods, and not eating or eating too much of them. Neither way will help.

  You can only eat three meals a day. Similarly, you can ONLY absorb a certain amount of information during your exam prep. So, focus on the most important stuff.

  Focus on your weaknesses but still read the other info. The key is to understand, digest the materials, and retain the information.

  It is NOT how much you have read, but how much you understand, digest, and retain that counts.

  The key to passing an ARE exam, or any other exam, is to know the scope of the exam, and not to read too many books. Select one or two really good books and focus on them. Actually <u>understand</u> the content and <u>memorize</u> it. For your convenience, I have <u>underlined</u> the fundamental information that I think is very important. You definitely need to <u>memorize</u> all the information that I have underlined.

  You should try to understand the content first, and then memorize the content of the book by reading it multiple times. This is a much better way than relying on "mechanical" memory without understanding.

When you read the materials, ALWAYS keep the following in mind:

- **Think like an architect.**
  For example, when you take the SS ARE exam, focus on what need to know to be able to coordinate your structural engineer's work, or tell them what to do. You are NOT taking an exam for becoming a structural engineer; you are taking an exam to become an architect.

  This criterion will help you filter out the materials that are irrelevant, and focus on the right information. You will know what to flip through, what to read line by line, and what to read multiple times.

  I have said this one thousand times, and I will say it once more:
  Time and effort is the most valuable asset of a candidate. How to cherish and effectively use your limited time and effort is the key to passing any exam.

15. ARE exam preparation requires short-term memory

You should understand that ARE Exam Preparation requires **Short-Term Memory**. This is especially true for the MC portion of the exam. You should schedule your time accordingly: in the early stages of your ARE exam Preparation, you should focus on understanding and an **initial** review of the material; in the late stages of your exam preparation, you should focus on memorizing the material as a **final** review.

16. Allocation of your time and scheduling

You should spend about 60% of your effort on the most important and fundamental study materials, about 30% of your effort on mock exams, and the remaining 10% on improving your weakest areas, i.e., reading and reviewing the questions that you answered incorrectly, reinforcing the portions that you have a hard time memorizing, etc.

Do NOT spend too much time looking for obscure ARE information because the NCARB will HAVE to test you on the most **common** architectural knowledge and information. At least 80% to 90% of the exam content will have to be the most common, important and fundamental knowledge. The exam writers can word their questions to be tricky or confusing, but they have to limit themselves to the important content; otherwise, their tests will NOT be legally defensible. At most, 10% of their test content can be obscure information. You only need to answer about 60% of all the questions correctly. So, if you master the common ARE knowledge (applicable to 90% of the questions) and use the guess technique for the remaining 10% of the questions on the obscure ARE content, you will do well and pass the exam.

On the other hand, if you focus on the obscure ARE knowledge, you may answer the entire 10% obscure portion of the exam correctly, but only answer half of the remaining 90% of the common ARE knowledge questions correctly, and you will fail the exam. That is why we have seen many smart people who can answer very difficult ARE questions correctly because they are able to look them up and do quality research. However, they often end up failing ARE exams because they cannot memorize the common ARE knowledge needed on the day of the exam. ARE exams are NOT an open-book exams, and you cannot look up information during the exam.

The **process of memorization** is like **filling a cup with a hole at the bottom**: You need to fill it faster than the water leaks out at the bottom, and you need to constantly fill it; otherwise, it will quickly be empty.

Once you memorize something, your brain has already started the process of forgetting it. It is natural. That is how we have enough space left in our brain to remember the really important things.

It is tough to fight against your brain's natural tendency to forget things. Acknowledging this truth and the fact that you cannot memorize everything you read, you need to focus your limited time, energy and brainpower on the most important issues.

The biggest danger for most people is that they memorize the information in the early stages of their exam preparation, but forget it before or on the day of the exam and still THINK they remember them.

Most people fail the exam NOT because they cannot answer the few “advanced” questions on the exam, but because they have read the information but can NOT recall it on the day of the exam. They spend too much time preparing for the exam, drag the preparation process on too long, seek too much information, go to too many websites, do too many practice questions and too many mock exams (one or two sets of mock exams can be good for you), and **spread themselves too thin**. They end up **missing the most important information** of the exam, and they will fail.

The ARE Mock Exam series along with the tips and methodology in each of the books will help you find and improvement your weakness areas, MEMORIZE the most important aspects of the test to pass the exam ON THE FIRST TRY.

So, if you have a lot of time to prepare for the ARE exams, you should plan your effort accordingly. You want your ARE knowledge to peak at the time of the exam, not before or after.

For example, if you have two months to prepare for a very difficult ARE exam, you may want to spend the first month focused on reading and understanding all of the study materials you can find as your **initial** review. Also during this first month, you can start memorizing after you understand the materials as long as you know you HAVE to review the materials again later to retain them. If you have memorized something once, it is easier to memorize it again later.

Next, you can spend two weeks focused on memorizing the material. You need to review the material at least three times. You can then spend one week on mock exams. The last week before the exam, focus on retaining your knowledge and reinforcing your weakest areas. Read the mistakes that you have made and think about how to avoid them during the real exam. Set aside a mock exam that you have not taken and take it seven days before test day. This will alert you to your weaknesses and provide direction for the remainder of your studies.

If you have one week to prepare for the exam, you can spend two days reading and understanding the study material, two days repeating and memorizing the material, two days on mock exams, and one day retaining the knowledge and enforcing your weakest areas.

The last one to two weeks before an exam is absolutely critical. You need to have the "do or die" mentality and be ready to study hard to pass the exam on your first try. That is how some people are able to pass an ARE exam with only one week of preparation.

17. Timing of review: the 3016 rule; memorization methods, tips, suggestions, and mnemonics

Another important strategy is to review the material in a timely manner. Some people say that the best time to review material is between 30 minutes and 16 hours (the **3016** rule) after you read it for the first time. So, if you review the material right after you read it for the first time, the review may not be helpful.

I have personally found this method extremely beneficial. The best way for me to memorize study materials is to review what I learn during the day again in the evening. This, of course, happens to fall within the timing range mentioned above.

Now that you know the **3016** rule, you may want to schedule your review accordingly. For example, you may want to read new study materials in the morning and afternoon, then after dinner do an initial review of what you learned during the day.

OR

If you are working full time, you can read new study materials in the evening or at night and then get up early the next morning to spend one or two hours on an initial review of what you learned the night before.

The initial review and memorization will make your final review and memorization much easier.

Mnemonics is a very good way for you to memorize facts and data that are otherwise very hard to memorize. It is often arbitrary or illogical but it works.

A good mnemonic can help you remember something for a long time or even a lifetime after reading it just once. Without the mnemonics, you may read the same thing many times and still not be able to memorize it.

There are a few common Mnemonics:

1) **Visual** Mnemonics: Link what you want to memorize to a visual image.
2) **Spatial** Mnemonics: link what you want to memorize to a space, and the order of things in it.
3) **Group** Mnemonics: Break up a difficult piece into several smaller and more manageable groups or sets, and memorize the sets and their order. One example is the grouping of the 10 digit phone number into three groups in the U.S. This makes the number much easier to memorize.
4) **Architectural** Mnemonics: A combination of Visual Mnemonics and Spatial Mnemonics and Group Mnemonics.

Imagine you are walking through a building several times, along the same path. You should be able to remember the order of each room. You can then break up the information that you want to remember and link them to several images, and then imagine you hang the images on walls of various rooms. You should be able to easily recall each group in an orderly manner by imagining you are walking through the building again on the same path, and looking at the images hanging on walls of each room. When you look at the images on the wall, you can easily recall the related information.

You can use your home, office or another building that you are familiar with to build an Architectural Mnemonics to help you to organize the things you need to memorize.

5) **Association** Mnemonics: You can associate what you want to memorize with a sentence, a similarly pronounced word, or a place you are familiar with, etc.
6) **Emotion** Mnemonics: Use emotion to fix an image in your memory.
7) **First Letter** Mnemonics: You can use the first letter of what you want to memorize to construct a sentence or acronym. For example, **"Roy G. Biv"** can be use to memorize the order of the 7 colors of the rainbow, it is composed of the first letter of each primary color.

You can use **Association** Mnemonics and memorize them as all the plumbing fixtures for a typical home, PLUS Urinal.

OR

You can use "Water S K U L" (**First Letter** Mnemonics selected from website below) to memorize them:

Water Closets
Shower
Kitchen Sinks
Urinal
Lavatory

18. The importance of good and effective study methods

There is a saying: Give a man a fish, feed him for a day. Teach a man to fish, feed him for a lifetime. I think there is some truth to this. Similarly, it is better to teach someone HOW to study than just give him good study materials. In this book, I give you good study materials to save you time, but more importantly, I want to teach you effective study methods so that you can not only study and pass ARE exams, but also so that you will benefit throughout the rest of your life for anything else you need to study or achieve. For example, I give you samples of mnemonics, but I also teach you the more important thing: HOW to make mnemonics.

Often in the same class, all the students study almost the SAME materials, but there are some students that always manage to stay at the top of the class and get good grades on exams. Why? One very important factor is they have good study methods.

Hard work is important, but it needs to be combined with effective study methods. I think people need to work hard AND work SMART to be successful at their work, career, or anything else they are pursuing.

19. The importance of repetition: read this book <u>at least</u> three times

Repetition is one of the most important tips for learning. That is why I have listed it under a separate title. For example, you should treat this book as part of the core study materials for your ARE exams and you need to read this book <u>at least three times</u> to get all of its benefits:

1) The first time you read it, it is new information. You should focus on understanding and digesting the materials, and also do an <u>initial</u> review with the **3016** rule.
2) The second time you read it, focus on reading the parts <u>I</u> have already highlighted AND <u>you</u> have <u>highlighted</u> (the important parts and the weakest parts for you).
3) The third time, focus on <u>memorizing</u> the information.

Remember the analogy of the <u>memorization process</u> as **filling a cup with a hole on the bottom**?
Do NOT stop reading this book until you pass the real exam.

20. The importance of a routine

A routine is very important for studying. You should try to set up a routine that works for you. First, look at how much time you have to prepare for the exam, and then adjust your current routine to include exam preparation. Once you set up the routine, stick with it.

For example, you can spend from 8:00 a.m. to 12:00 noon, and 1:00 p.m. to 5:00 p.m. on studying new materials, and 7:00 p.m. to 10:00 p.m. to do an initial review of what you learned during the daytime. Then, switch your study content to mock exams, memorization and retention when it gets close to the exam date. This way, you have 11 hours for exam preparation everyday. You can probably pass an ARE exam in one week with this method. Just keep repeating it as a way to <u>retain</u> the architectural knowledge.

OR
You can spend 7:00 p.m. to 10:00 p.m. on studying new materials, and 6:00 a.m. to 7:00 a.m. to do an initial review of what you learned the evening before. This way, you have four hours for exam preparation every day. You can probably pass an ARE exam in two weeks with this preparation schedule.

A routine can help you to memorize important information because it makes it easier for you to concentrate and work with your body clock.

Do NOT become panicked and change your routine as the exam date gets closer. It will not help to change your routine and pull all-nighters right before the exam. In fact, if you pull an all-nighter the night before the exam, you may do much worse than you would have done if you kept your routine.

All-nighters or staying up late are not effective. For example, if you break your routine and stay up one-hour late, you will feel tired the next day. You may even have to sleep a few more hours the next day, adversely affecting your study regimen.

21. The importance of short, frequent breaks and physical exercise

Short, frequent breaks and physical exercise are VERY important for you, especially when you are spending a lot of time studying. They help relax your body and mind, making it much easier for you to concentrate when you study. They make you more efficient.

Take a five-minute break, such as a walk, at least once every one to two hours. Do at least 30 minutes of physical exercise every day.

If you feel tired and cannot concentrate, stop, go outside, and take a five-minute walk. You will feel much better when you come back.

You need your body and brain to work well to be effective with your studying. Take good care of them. You need them to be well-maintained and in excellent condition. You need to be able to count on them when you need them.

If you do not feel like studying, maybe you can start a little bit on your studies. Just casually read a few pages. Very soon, your body and mind will warm up and you will get into study mode.

Find a room where you will NOT be disturbed when you study. A good study environment is essential for concentration.

22. A strong vision and a clear goal

You need to have a strong vision and a clear goal: to master the architectural knowledge and become an architect in the shortest time. This is your number one priority. You need to master the architectural knowledge BEFORE you do sample questions or mock exams, except "truth-finding" exam or exercise at the very beginning of your exam prep. It will make the process much easier. Everything we discuss is to help you achieve this goal.

As I have mentioned on many occasions, and I say it one more time here because it is so important:

It is how much architectural knowledge and information you can understand, digest, memorize, and firmly retain that matters, not how many books you read or how many sample tests you have taken. The books and sample tests will NOT help you if you cannot understand, digest, memorize, and retain the important information for the ARE exam.

Cherish your limited time and effort and focus on the most important information.

23. English system (English or inch-pound units) vs. metric system (SI units)

This book is based on the English system or English units. The English or inch-pound units are based on the module used in the U.S. Effective July 2013, the ARE includes measurements in inch-pound units only. Metric system (SI units) is no longer used.

24. Codes and standards used in this book

We use the following codes and standards:
American Institute of Architects, Contract Documents, Washington, DC.
Canadian Construction Documents Committee, CCDC Standard Documents, 2006, Ottawa.

25. Where can I find study materials on architectural history?

Every ARE exam may have a few questions related to architectural history. The following are some helpful links to FREE study materials on the topic:

http://www.essentialhumanities.net/arch.php
http://issuu.com/motimar/docs/history_synopsis?viewMode=magazine
http://www.ironwarrior.org/ARE/Materials_Methods/m_m_notes_2.pdf

Chapter Two

California Supplemental Exam (CSE)

A. First Thing First: Go to the Website of the California Architect Board (CAB) and Read all the Requirements for the California Supplemental Exam.
See the following link:
http://www.cab.ca.gov/candidates/cse.shtml

B. Download and Review the Latest CSE Documents from the CAB Website.

We'll walk you through each of the CSE documents listed on the CAB website and tell you what to look for in each document, what documents to read multiple times, and what documents to simply read through once.

1. A detailed list and brief description of the FREE PDF files and information that you can download from the CAB website

The following is a detailed list of the FREE PDF files or information that you can download from the CAB website.

- **CSE Overview** gives you a basic introduction to the CSE. Skimming through this should be adequate.

  Note: *If you fail the CSE, you need to wait 180 days before submitting a new application and payment to CAB in order to retake the exam.*

- **Eligibility and Scheduling**
  The requirements to qualify for the CSE are as follows:
  a. eight years of post-secondary education and/or work experience as evaluated by the Board (including at least one year of work experience under the direct supervision of an architect licensed in a U.S. jurisdiction or two years of work experience under the direct supervision of an architect registered in a Canadian province)
  b. completion of the Intern Development Program (IDP)
  c. successful completion of the Architect Registration Examination (ARE)

  California does NOT require an accredited degree in architecture for examination and licensure. However, many other states do.

  Note: *CAB's examination vendor for the CSE is Psychological Services, LLC (PSI) -www.psiexams.com.*

A candidate can take the CSE at 17 PSI examination sites in California and 22 additional sites out of state. Examinations may be scheduled during normal working hours of 8:00 a.m. to 5:00 p.m., Monday through Friday, and operating hours on Saturday.

- **Exam Security and Confidentiality** includes a list of disciplinary actions for divulging ARE and CSE exam content per Section 123 and Section 496 of the Business and Professions Code:
  § 123 Subversion of Licensing Examinations – Misdemeanor
  § 123.5 Engagement in Practices Constituting a Violation Under § 123; Injunction or Restraining Order
  § 496 Denial, Suspension; or Violation of § 123; Revocation of License
  (Skimming through this should be adequate.)

- **Preparing for the CSE**
  CAB suggests the following:
  a. Thoroughly read the CSE Handbook.
  b. Study the CSE Test Plan.
  c. Use the CSE reference materials.
  d. Engage in self-directed study.
  e. Enroll in examination seminars or use independent study guides.

- **CSE Handbook** includes information on administrative and exam procedures, etc.

  See the FREE PDF file at the following link:
  http://www.cab.ca.gov/pdf/publications/cse_handbook.pdf

- **CSE Test Plan** defines the scope of the CSE and includes four areas of architectural knowledge and abilities, and 55 knowledge and ability statements. This is the most important document, and EVERY question of the CSE is based on this plan. You need to read this at least three times. The CSE does not duplicate the general areas of architectural practice and knowledge already covered by the ARE, but instead focuses on California-specific areas of practice.

  See the following link:
  http://www.cab.ca.gov/candidates/cse/test_plan.shtml

- **Taking the CSE** provides specific introduction on CSE content. The CSE includes *approximately* 100 multiple-choice items. There is only one correct answer for each question, and no point deduction for incorrect answers. So, you should make sure you answer every question. If you do not know the answer, you should put down a guess answer to increase your chance of passing the CSE.

  The CSE has *two* separate timed sections with a total time of 3.5 hours. If you finish the first section early, you can NOT use the extra time for the second section.

CSE includes the following two sections:

1) **Project Scenario Section** has multiple-choice items based on a hypothetical project. Make sure your paper handouts match the project on your screen. If not, you need to let the proctor know right away. Do not spend too much time reading the drawings, and only refer to the drawings when the question asks for information on the drawings.
2) **General Section** contains general multiple-choice items based on content within the CSE Test Plan

- **Reasonable Accommodations** provides information for individuals with disabilities. "Requests for accommodation must be received a minimum of 90 days prior to the desired test date to allow for processing."

- **Examination Results** are provided to candidates at examination sites upon completion of their examination. Candidates can file for an appeal or review with the CAB within 30 days, but a score can never be reversed.

- **Occupational Analysis Study 2007 (Summary)** is the base for current version of the CSE and CSE Test Plan, and includes many pages of survey questions. This is an important document. Skimming through the document should be adequate since it has already been incorporated into the CSE Test Plan.

  Based on this survey, CAB develops *four* weighted primary knowledge categories:

Categories and Sub-Categories	**Weight**
1) Context and Pre-Design	16%
2) Regulatory	42%
a) California State Laws, Codes, Regulations, and Standards	24%
b) Other Laws, Codes, Regulations, Standards, Agencies, and Entitlements	18%
3) Management and Design	27%
4) Construction	15%

C. CSE Reference Materials (Official CAB Reference Materials *Plus* Additional Critical Materials)

See the latest information and a list of official CAB reference materials at the following link:
http://www.cab.ca.gov/candidates/cse_materials.shtml

In addition to the official CAB reference materials, we have added some other very important materials that will be tested. Let us go over these reference materials in detail.

1. **American Institute of Architects (AIA) Documents**
   You may have some real CSE questions based on AIA documents.
   Reading the summary of the AIA Documents is NOT adequate preparation. You need to read the complete text. Fortunately, you do NOT have to read all the available AIA documents.

   Three possible study solutions are as follows:

a. Buy ONLY the AIA documents you need from your local AIA office. The AIA documents listed below are important, read them at least three times. You may have some real CSE questions based on the following AIA documents:

- **A201–2007, General Conditions of the Contract for Construction**
- **B102–2007, Standard Form of Agreement Between Owner and Architect without a Pre-defined Scope of Architect's Service**

  You can find FREE sample forms and commentaries for AIA document A201 at the following link:
  http://www.aia.org/contractdocs/aiab081438

  You can find FREE sample form for AIA document B102 at the following links:
  http://www.aia.org/aiaucmp/groups/aia/documents/pdf/aiab093936.pdf
  http://www.aia.org/contractdocs/aiab081438

- **B201–2007, Standard Form of Architect's Services: Design and Construction Contract Administration**

  You can find FREE sample form for AIA document documents B201 at the following links:
  http://www.aia.org/contractdocs/aiab081438

b. Buy Architect's Handbook of Professional Practice (AHPP) which has a CD including the sample AIA documents. AHPP itself is also a very important publication for the PPP division.

c. After you register with NCARB for the ARE exams and log in to their site, you have FREE access to the AIA documents online.

2. **California Architects Practice Act (Latest Edition)**
   Reading through this twice should be adequate.

   See the following link:
   http://www.cab.ca.gov/pdf/publications/architects_practice_act_2012_march.pdf

3. **California Contractors License Law and Reference Book (Lien Laws)**
   Skimming through these should be adequate.

   You can get them at a local county law library, or from the Contractors State License Board (CSLB) website at the following link:
   http://www.cslb.ca.gov/GeneralInformation/Library/Laws/

4. **California Public Contract Code (State Contract Act 10100 - 10285.5)**
   Skimming through this should be adequate.

   You can get this at a local county law library, or from the Leginfo website at the following link:
   http://www.leginfo.ca.gov/calaw.html

5. **General Plan (Government Code, Section 65300 Et. Seq.)**
   Skimming through this should be adequate.

   You can get this at a local county law library, or at the following links:
   http://codes.lp.findlaw.com/cacode/GOV/1/7/d1/3/5/s65300
   http://www.leginfo.ca.gov/calaw.html

6. **Essential Services Buildings Seismic Safety Act, Excerpt (California Health & Safety Code, Sections 16000 - 16023)**
   Skimming through this should be adequate.

   You can get this at a local county law library, or at the following links:
   http://www.cab.ca.gov/licensees/essential_services_act.shtml
   http://codes.lp.findlaw.com/cacode/HSC/1/d12.5/2
   http://www.leginfo.ca.gov/calaw.html

7. **Hospital Facilities Seismic Safety Act (HFSSA)**

   See the following links:
   http://www.seismic.ca.gov/pub/CSSC_2001-04_Hospital.pdf
   http://www.oshpd.ca.gov/FDD/seismic_compliance/SB1953/index.html

8. **California Design Professionals' Liens (California Civil Code, Sections 8300 - 8319)**
Skimming through this should be adequate.
Design professional liens are gone but not forgotten. See the following link:
http://www.aiacc.org/2012/02/27/design-professional-liens-gone-but-not-forgotten/

Skimming through this should be adequate.

9. **California Coastal Act (Public Resources Code, Section 30000 Et. Seq.)**
Skimming through this should be adequate.

You can get this from the California Coastal Commission, or at the following links:
http://www.coastal.ca.gov/coastact.pdf
http://www.leginfo.ca.gov/calaw.html

10. **Architect's Handbook of Professional Practice**
Skimming through this and focusing on the portions that you are not familiar with should be adequate.

Demkin, Joseph A., AIA, Executive Editor. *The Architect's Handbook of Professional Practice* (AHPP). The American Institute of Architects & Wiley, latest edition.

This comprehensive book covers all aspects of architectural practice, and includes two CDs containing sample AIA contract documents.

You can purchase this from American Institute of Architects or American Institute of Architects California Council, or at the following link:
http://www.amazon.com/

11. **California Building Standards Code by California Building Standards Commission**
Skimming through these and focusing on the portions that you are not familiar with should be adequate. Focus on the portions that you as an architect will frequently use, such as exit width calculations, occupancy groups, lateral forces, California handicap requirements, plumbing fixture calculations, sizing of roof drains, etc. See my book, *Building Construction*, **ISBN-13:** 9780984374144, for more information.

California Building Standards Code is also called Title 24 of the California Code of Regulations (CCR).
http://www.bsc.ca.gov/pubs/codeson.aspx
http://www.dgs.ca.gov/dsa/Programs/progCodes/title24.aspx

Note:
These codes will be updated every three years. You just need to find the most current version of the codes at these links.

One way to arrange the codes is to divide Title 24 into 12 parts:

- Part 1 - California Building Standards Administrative Code
- Part 2 - California Building Code
- (Part 2.5 - California Residential Building Code)
- Part 3 - California Electrical Code
- Part 4 - California Mechanical Code
- Part 5 - California Plumbing Code
- Part 6 - California Energy Code
- Part 7 - California Elevator Safety Construction Code
- Part 8 - California Historical Building Code
- Part 9 - California Fire Code
- Part 10 - California Existing Building Code
- Part 11 - California Green Building Standards Code
- Part 12 - California Reference Standards Code

For those of you who do not have a copy of various California Building Codes handy, the following information may help:

The California Building Codes are based on the International Building Codes, and most of the information is the same. California is subject to many earthquakes. So, pay attention to any information related to earthquakes.

You can find various California Building Codes for FREE at the following website:
http://publicecodes.cyberregs.com/st/ca/st/index.htm

- ***2013* California *Building Code (IBC)***

Focus on **Chapter 16**, particularly the sections on earthquakes and wind. Read them a few times, and have a general idea of the concepts. Do not force yourself to memorize all the details.

Read the following pages and become familiar with the Uniform and Concentrated Loads CBC table 1607.6.

You can view the Uniform and Concentrated Loads IBC Table 1607.1 for FREE at the following link:
http://publicecodes.cyberregs.com/st/ca/st/b200v10/st_ca_st_b200v10_16_par087.htm

Note:

- *If the codes are updated in the future, you just need to go to the root directory and find the latest version of the codes:*
  http://publicecodes.cyberregs.com

- *You need to spend a large percentage of your prep time on CBC Chapter 16 and become familiar with it. You do not need to force yourself to memorize the numbers and all the details. Just reading it a few times and becoming familiar with the information should be adequate.*

12. California Health and Safety Code (CHSC)
Skimming through them and focusing on the portions related to design and construction should be adequate.

See the following link:
http://www.leginfo.ca.gov/.html/hsc_table_of_contents.html

13. Mechanics Lien
Skimming through this should be adequate.

See the following link:
http://www.cslb.ca.gov/Consumers/LegalIssuesForConsumers/MechanicsLien/UnderstandingMechanicsLiens.asp

14. California Environmental Quality Act or CEQA (Public Resources Code, Section 21000 Et. Seq.)
Reading through it and focusing on the portions that you are not familiar with should be adequate.

You can get the document from American Council of Engineering Companies, or California Environmental Resources Evaluation System, or at the following link:
http://www.opr.ca.gov/m_ceqa.php

15. Information on the Natural Environment in California

- **Wetlands**
  See the following link:
  http://ceres.ca.gov/ceres/calweb/wetlands.html

- **Coastal California**
  See the following link:
  http://en.wikipedia.org/wiki/Coastal_California

- **California Endangered Species Act (CESA)**
  See the following link:
  http://www.dfg.ca.gov/habcon/cesa/

16. California Coastal Zone Conservation Act (Public Resources Code, Division 20)
Reading through it and focusing on the portions that you are not familiar with should be adequate.

You can get it at the following link:
http://www.coastal.ca.gov/coastact.pdf

17. California Clean Air Act
Skimming through this should be adequate.

See the following link:
http://www.calepa.ca.gov/Legislation/1996/ab3048.htm

18. The Field Act
Reading through it and focusing on the portions that you are not familiar with should be adequate.

See the following link:
http://en.wikipedia.org/wiki/Field_Act

19. State of California Indoor Air Quality Program
Reading through it and focusing on the portions that you are not familiar with should be adequate.

You can get it at the following links:
http://www.cdph.ca.gov/programs/IAQ/Pages/default.aspx
http://www.epa.gov/iaq/states/california.html

20. Minimum Design Loads for Buildings and Other Structures, 2005 ASCE 7-05
(Skimming through them and focus on the portions that you are not familiar with should be adequate.)

ASCE 7 is a book entitled *Minimum Design Loads for Buildings and Other Structures*, published by ASCE (American Society of Civil Engineers). You can purchase it from Amazon, but I doubt you'll refer to it a lot in practice, because you normally leave the detailed structural calculations to your structural engineer.

Per ASCE 7, 1609.4 & 6.5.6 Exposure Categories are defined as follows.
- **Exposure A** is no longer used in ASCE 7.
- **Exposure B** is used as the default.
- **Exposure C** includes shorelines of hurricane prone regions (no longer Exposure D).
- **Exposure D** now only applies to inland waterways, Great Lakes, Coastal California, Oregon, Washington, and Alaska.

See page 20 of the PDF file of "Wind Design Made Simple" by ICC TRI-Chapter Uniform Code Committee available for FREE at the following link:
http://www.calbo.org/Documents/SimplifiedWindHandout.pdf

IBC also has similar definitions.

"An exposure category shall be determined in accordance with the following:

Exposure B. Exposure B shall apply where the ground surface roughness condition, as defined by Surface Roughness B, prevails in the upwind direction for a distance of at least 2,600 feet (792 m) or 20 times the height of the building, whichever is greater.

Exception: For buildings whose mean roof height is less than or equal to 30 feet (9144 mm), the upwind distance is permitted to be reduced to 1,500 feet (457 m).

Exposure C. Exposure C shall apply for all cases where Exposures B or D do not apply.

Exposure D. Exposure D shall apply where the ground surface roughness, as defined by Surface Roughness D, prevails in the upwind direction for a distance of at least 5,000 feet (1524 m) or 20 times the height of the building, whichever is greater. Exposure D shall extend inland from the shoreline for a distance of 600 feet (183 m) or 20 times the height of the building, whichever is greater."

See related IBC information for FREE at the following link:
http://publicecodes.cyberregs.com/icod/ibc/2006f2/icod_ibc_2006f2_16_par100.htm

21. **Local Government Regulations (i.e., design review standards, development project regulations, and/or ordinances)**
    Skimming through them and focusing on the portions that you are not familiar with should be adequate.

    You can get them from various local sources.

22. **The FREE PDF file of FEMA publication number 454 (FEMA454),** *Designing for Earthquakes: A Manual for Architects*, is available at the following link:
    http://www.fema.gov/library/viewRecord.do?id=2418

    Note:
    Earthquake and lateral forces are major concerns for California Architects. You need to spend a large percentage (at least 30%) of your prep time on this ***FEMA454*** *PDF file. Focus on* ***Chapters 4, 5, 8, and 9****. You may have* ***many real CSE questions*** *based on these chapters. This PDF book has many diagrams and photos and helps you understand what happened in buildings that failed during an earthquake.*

23. **The FREE PDF file of *Wind Design Made Simple*** by ICC TRI-Chapter Uniform Code Committee is available at the following link:
    http://www.calbo.org/Documents/SimplifiedWindHandout.pdf

Note:
You need to become familiar with this file. You do not need to force yourself to memorize the numbers and all the details, just reading it a few times and becoming familiar with the information should be adequate.

24. *Buildings at Risk* by Christopher Arnold
Reading through this twice should be adequate. You can find this book by a Google search with its title and author.

25. The FREE PDF file of *Non-structural Earthquake Hazards in California Schools*
Reading through this twice should be adequate.

The FREE PDF file is available at the following link:
http://www.documents.dgs.ca.gov/dsa/pubs/SB1122.pdf

This document is applicable not only for schools, but also other kinds of buildings in California as well.

26. Construction Specifications Institute (CSI) MasterFormat & *Building Construction*
Become familiar with the new 6-digit CSI Construction Specifications Institute (CSI) MasterFormat as there may be a few questions based on this publication. Make sure you know which items/materials belong to which CSI MasterFormat specification section, and memorize the major section names and related numbers. For example, Division 9 is Finishes, and Division 5 is Metal, etc. Another one of my books, *Building Construction*, has detailed discussions on CSI MasterFormat specification sections.

Mnemonics for the 2004 CSI MasterFormat

The following is a good mnemonic, which relates to the 2004 CSI MasterFormat division names. Bold font signals the gaps in the numbering sequence.

This tool can save you lots of time: if you can remember the four sentences below, you can easily memorize the order of the 2004 CSI MasterFormat divisions. The number sequencing is a bit more difficult, but can be mastered if you remember the five bold words and numbers that are not sequential. Memorizing this material will not only help you in several divisions of the ARE, but also in real architectural practice

Mnemonics (pay attention to the underlined letters):
Good students can memorize material when teachers order.
F students earn F's simply 'cause **forgetting** principles have **an** effect. (21 and 25)
C students **end** everyday understanding things without memorizing. (31)
Please make professional pollution prevention inventions **everyday**. (40 and 48)

1-Good.................................. General Requirements
2-Students.............................. (Site) now Existing Conditions
3-Can..................................... Concrete
4-Memorize........................... Masonry
5-Material .... Metals
6-When.................................. Woods and Plastics
7-Teachers............................. Thermal and Moisture
8-Order.................................. Openings

9-F... Finishes
10-Students........................... Specialties
11-Earn.................................. Equipment
12-F's..................................... Furnishings
13-Simply.............................. Special Construction
14-'Cause............................... Conveying
21-Forgetting .... Fire
22-Principles......................... Plumbing
23-Have................................ HVAC
25-An.................................... Automation
26-Effect............................... Electric

27-C...................................... Communication
28-Students........................... Safety & Security
31-End.................................. Earthwork
32-Everyday.......................... Exterior
33-Understanding .... Utilities
34-Things.............................. Transportation
35-Without Memorizing........ Waterways and Marine

40-Please............................... Process Integration
41-Make................................ Material Processing and Handling Equipment
42-Professional..................... Process Heating, Cooling, and Drying Equipment
43-Pollution.......................... Process Gas and Liquid Handling, Purification and Storage Equipment
44-Prevention........................ Pollution Control Equipment
45-Inventions........................ Industry-Specific Manufacturing Equipment
48-Everyday......................... Electrical Power Generation

Note:
There are 49 CSI divisions. The "missing" divisions are those "reserved for future expansion" by CSI. They are intentionally omitted from the list.

D. Strategies and Tips for the CSE

Note:

While the majority of the MC questions in the real CSE exam ***focus on concepts and principles****, there may be* ***some tricky questions requiring extra time****. Therefore, it is absolutely necessary and critical for you to know how to handle them.*

In the CSE exams, it may be a good idea to skip any tricky question that requires over 30 seconds of your time; just pick a guess answer, mark it, and come back to it after completing the rest of the exam. This way, you have more time to read and answer other easier questions correctly.

A tricky question that takes 20 minutes to answer will gain the same number of points as a simple question that ONLY takes 2 minutes.

If you spend 20 minutes on a tricky question early on in the exam, you risk losing the time to read and answer ten easier subsequent questions, which could result in a loss of ten points instead of one.

Chapter Three

CSE Mock Exam

A. Mock Exam: Project Scenario Section

__Note:__ See the Project Plans on the following pages.

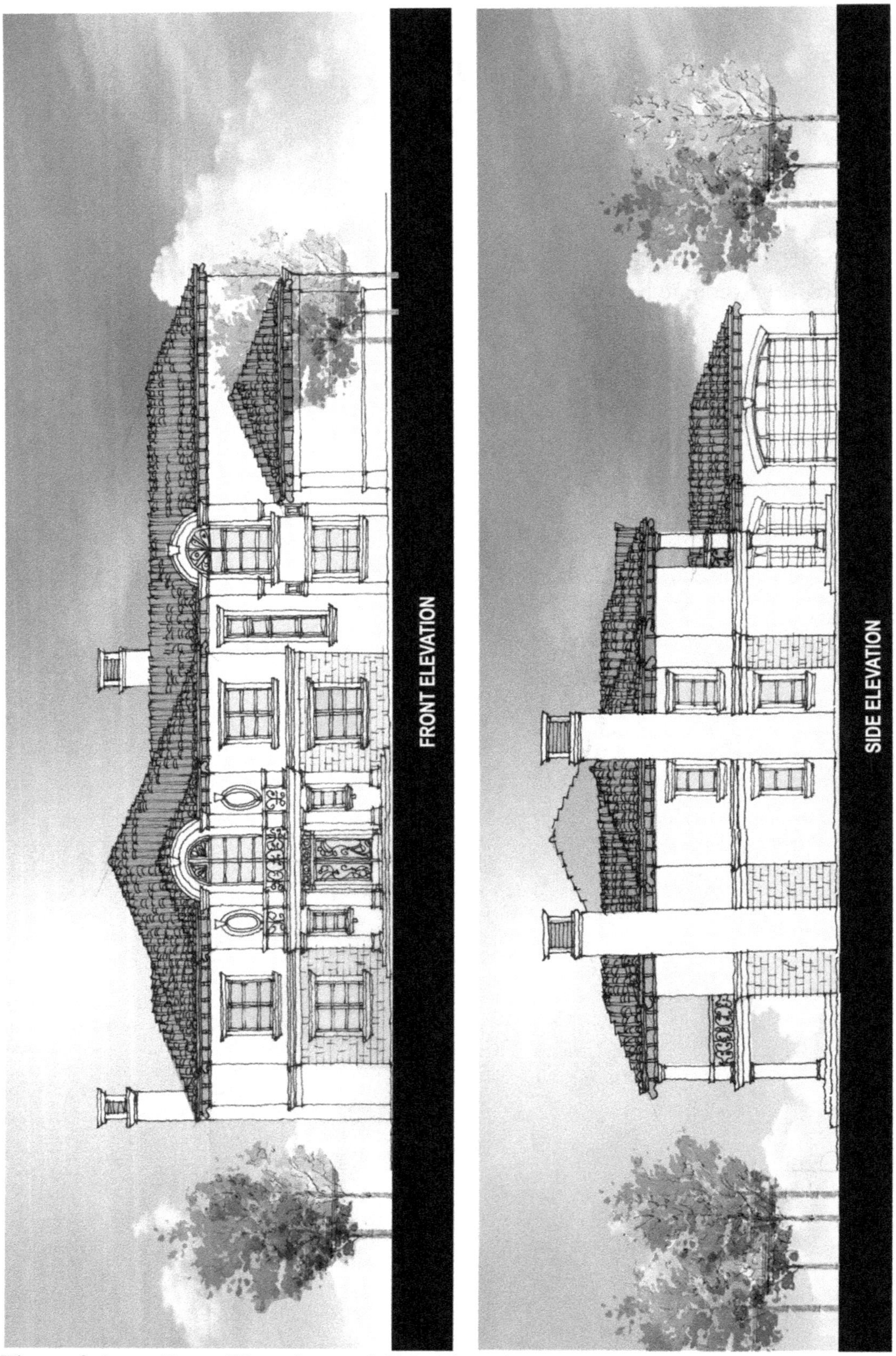

Figure 3.1 Front Elevation and Side Elevation

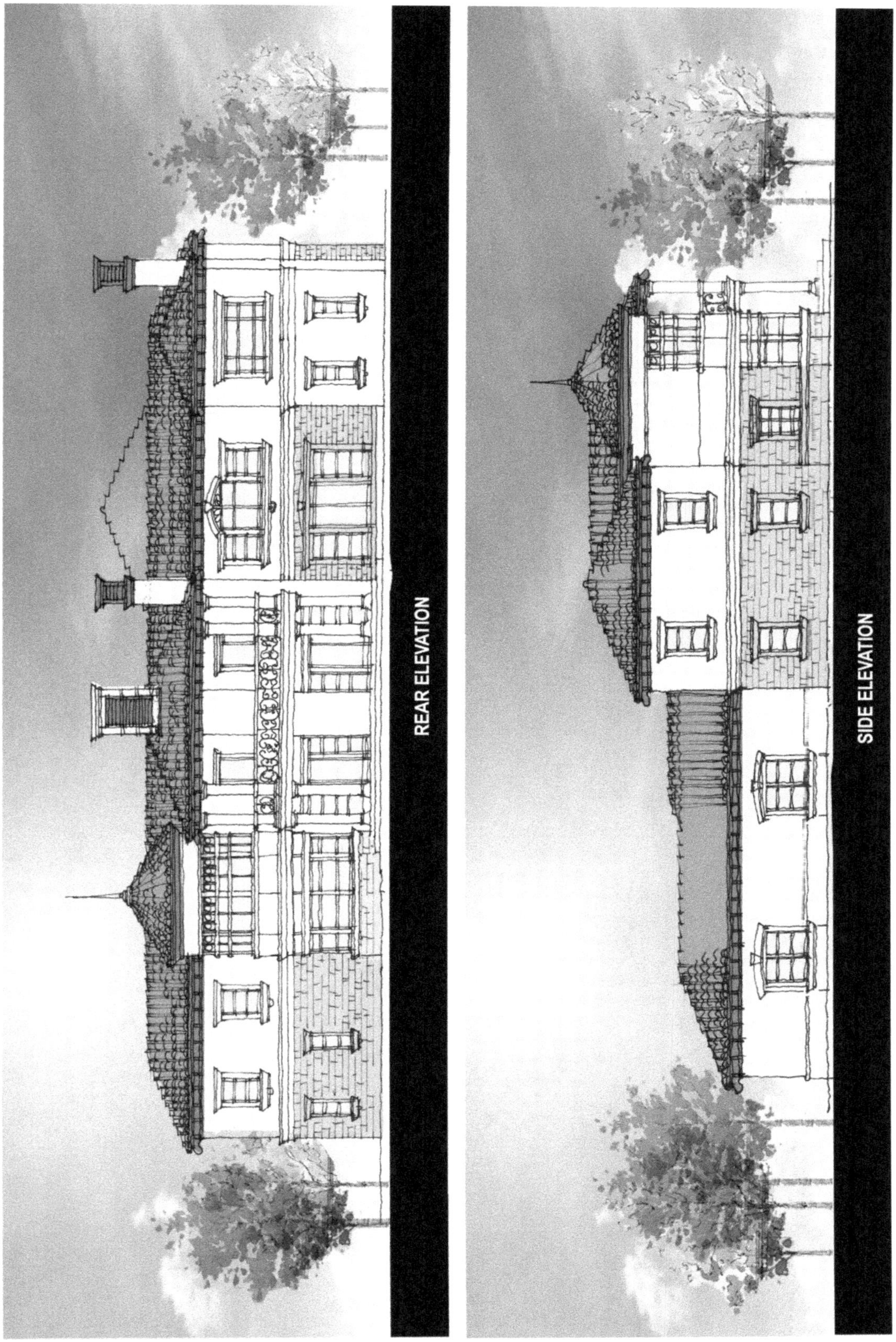

Figure 3.2 Rear Elevation and Side Elevation

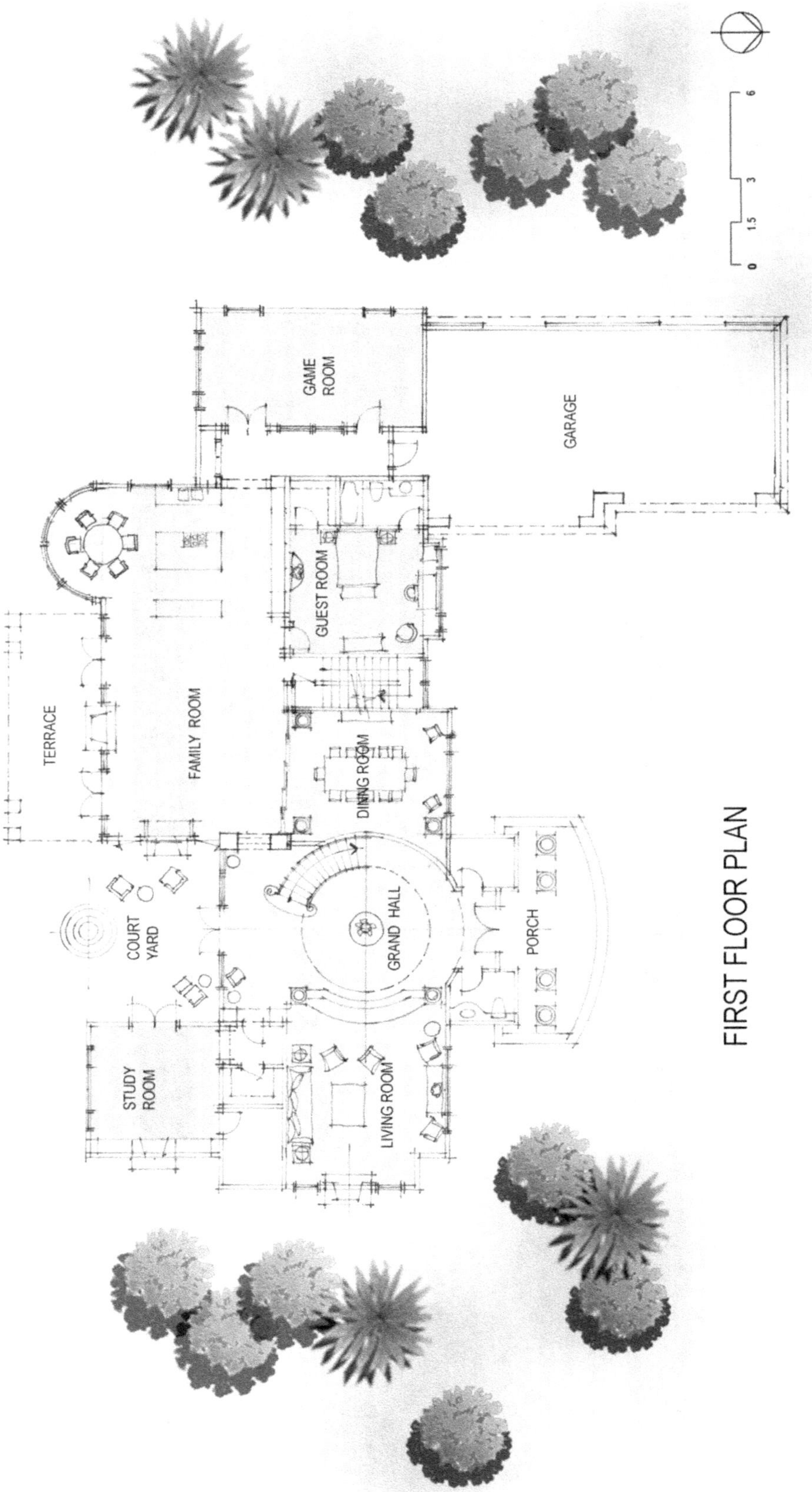

Figure 3.3 First Floor Plan

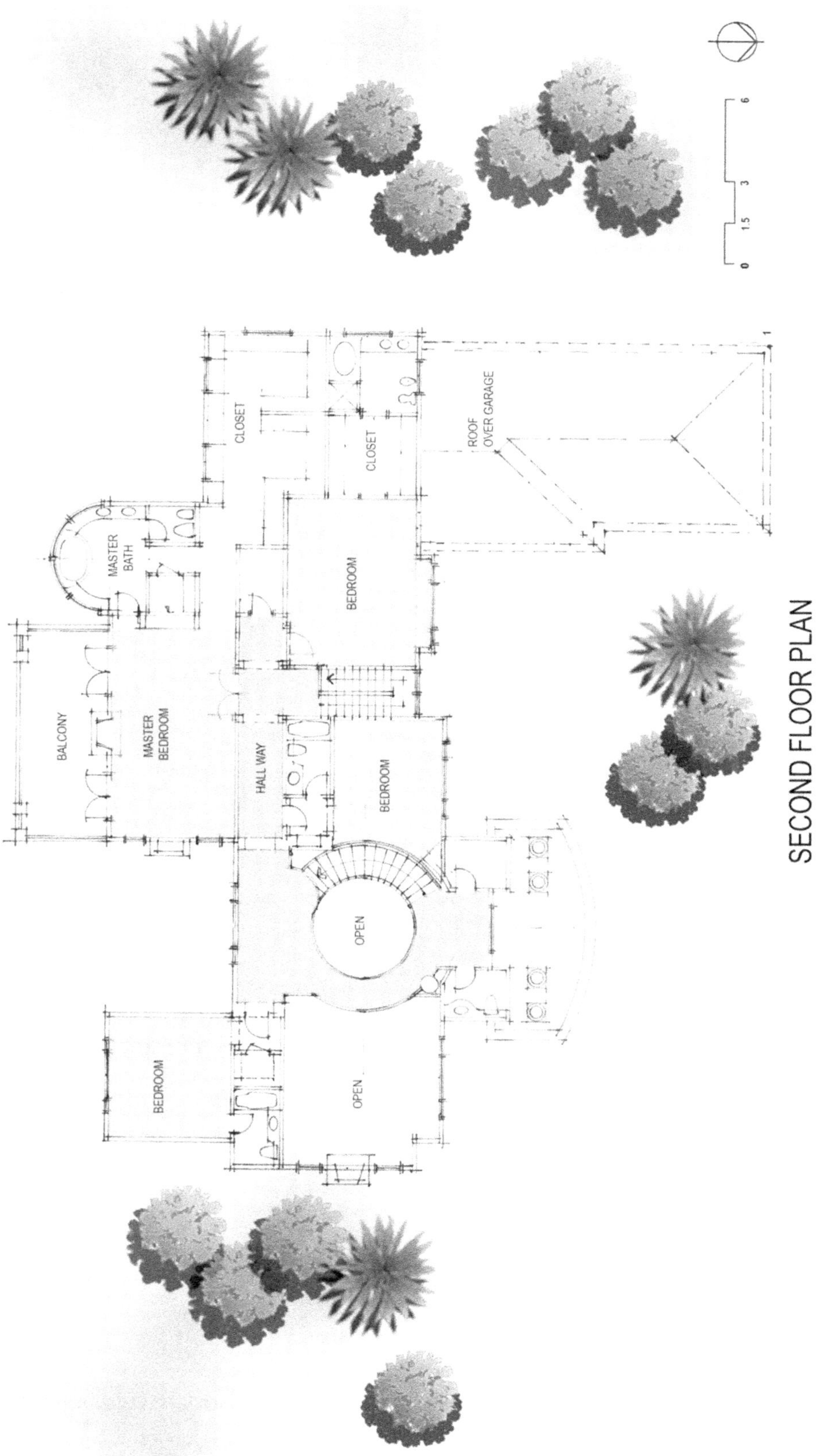

Figure 3.4 Second Floor Plan

Project Scenario

Project Site:

City of Newport Beach in Southern California
Two-acre site in a coastal zone
Fire hazard is high.
Seismic zone: 3
Next to Back Bay and within two miles of a small airport
Preliminary soils report shows bedrock at 25 feet, a high water table, and clay soil.

Project Scope:

The plans on previous pages show an existing two-story custom home. It has approximately 8,000 s.f., was built in 1938, and is listed on the city's registry of historical buildings. The owner donated the home to the county, and the county wants to convert the home into a public library, and add 3,800 s.f of new construction.

The building will be fully-sprinklered.

Program:

Front desk	800 s.f.
Reference desk (should be visible from the front desk)	1,000 s.f.
Children's reading area	1,200 s.f.
General reading area	1,500 s.f.
Public computer stations	1,000 s.f.
Bookshelves and isolated reading area	5,300 s.f.
Restrooms and supporting space	1,000 s.f.
Total Area	11,800 s.f

Architect's Profile:

ArchiteG, Inc. is an architectural firm in Irvine. The firm specializes in commercial, public, and residential projects. You are a principal of the firm.

Project Team:

Landscape Architect: Greenscape, Inc.
Structural Engineer: ABF, Inc.
Electrical Engineer: EGL, Inc.
Mechanical Engineer: MGI, Inc.
Plumbing Engineer: PCI, Inc.

Client's Profile:

The county has several departments involved, and the county supervisors have designated the Director of County Facilities Management as the point of contact.

Note: *Each question has only one correct answer.*

1. Based on the plans on previous pages, what is the style of the existing building?
   1) Colonial Style
   2) Spanish Style
   3) Mediterranean Style
   4) Arts and Crafts Style

2. Which occupancy group should the building be in for the purpose of code analysis?
   1) A
   2) B
   3) E
   4) R

3. Which of the following insulation is unlikely to be used in the exterior walls or roof of the new construction?
   1) R-11
   2) R-13
   3) R-19
   4) R-30

4. The following is part of CBC, Table 503 Allowable Building Heights and Areas. Based on the table, what is the most cost effective type of construction you should choose for this building for the purpose of code analysis?
   1) Type II-A
   2) Type III-A
   3) Type IV
   4) Type V-B

	TYPE OF CONSTRUCTION								
	TYPE I		TYPE II		TYPE III		TYPE IV	TYPE V	
	A	B	A	B	A	B	HT	A	B
HEIGHT (feet)	**UL**	**160**	**65**	**55**	**65**	**55**	**65**	**50**	**40**
	STORIES(S) AREA (A)								
S A	UL UL	11 UL	3 15,500	2 9,500	3 14,000	2 9,500	3 15,000	2 11,500	1 6,000

Table 3.1 Allowable Building Heights and Areas

5. Which of the following is the most cost effective way to comply with accessibility requirements?
   1) Add an elevator.
   2) Add a wheel chair lift.
   3) Provide a refuge area.
   4) Arrange the spaces to provide equal access on the first floor.

6. You are working on the project's detailed program. Which of the following users will you be unlikely to solicit information on spatial requirements from?
   1) library readers
   2) Director of Orange County Public Library (OCPL)
   3) Director of County Facilities Management
   4) none of the above

7. If there are conflicts between various county departments regarding the spatial requirements, whose instructions should you follow for this project?
   1) library readers
   2) Director of Orange County Public Library (OCPL)
   3) Director of County Facilities Management
   4) county supervisors

8. Which of the following may affect your project?
   1) HOA
   2) EIR
   3) fire department
   4) all of the above

9. Which of the following requires proper separation from other areas?
   1) reference desk
   2) children's reading area
   3) general reading area
   4) public computer stations

10. Which of the following should not have priority for the first floor spaces?
    1) front desk
    2) reference desk
    3) general reading area
    4) bookshelves and isolated reading area

11. In addition to schedule and budget, which of the following is the most important factor determining the feasibility of this project?
    1) availability of labors
    2) demography
    3) infrastructure
    4) zoning regulations

12. Which of the following is probably necessary for this project?
    1) CUP application
    2) variance
    3) eminent domain
    4) nonconforming use

13. Which of the following is not part of the UL business?
    1) product safety
    2) verification services
    3) environment
    4) none of the above

14. Per the California Coastal Act, which of the following actions can be permitted on wetlands?
    1) diking
    2) filling
    3) dredging
    4) all of the above
    5) none of the above

15. Per the California Coastal Act, who of the following has the designated authority to approve this project?
    1) City of Newport Beach Design Review Board
    2) City of Newport Beach Planning Department
    3) Californian Coastal Commission
    4) Orange County Board of Supervisors

16. Which of the following is a state agency that has jurisdiction over this project?
    1) Department of Fish and Game
    2) Division of the State Architect (DSA)
    3) Office of Statewide Health Planning and Development
    4) Water Quality Control Board

17. The planning department wants to have minimum lighting levels to avoid light pollution. The police department wants to have bright lighting levels to avoid crimes. What should you do as an architect?
    1) Follow the planning department's instructions to have minimum lighting levels since the planning department is in charge of the overall review of the project.
    2) Follow the police department's instructions to have bright lighting levels since avoiding crimes is a top priority.
    3) Ask the city's building official for a final decision.
    4) Work with both the planning department and the police department to come up with a compromise or a solution.

18. Who of the following can prepare a CLTA map with both topography and boundary information?
    1) a civil engineer
    2) a landscape architect
    3) a surveyor
    4) all of the above

19. The county has hired a civil engineer directly. The civil engineer has missed the project deadline several times, and delayed the architect's project schedule. What should the architect do?
    1) Contact the civil engineer directly.
    2) Contact the civil engineer and copy the county in all correspondence.
    3) Contact the county and let the county deal with the delay.
    4) none of the above

20. One of the mechanical ducts is too large and cannot fit between the ceiling and the structural members above. What is the most cost effective way for you to resolve this issue?
    1) Have the structural engineer revise the structural plans.
    2) Have the mechanical engineer revise the mechanical plans.
    3) Have the structural engineer and mechanical engineer communicate with each other in order to resolve this issue.
    4) Ask a contractor for suggestions.

21. What is an effective way to obtain stakeholder input?
    1) Hold a design charrette.
    2) Schedule a scope meeting with all departments of the city.
    3) Have a dinner party with all the stakeholders.
    4) Put an advertisement on local TV.

22. What is an effective way to ensure quality assurance in an architect's office?
    1) Hold a design charrette.
    2) Have a manufacturer's representative give a lunch seminar on a product used in the project.
    3) Keep a checklist.
    4) all of the above

23. What is the most effective way to ensure quality assurance for consultants' plans?
    1) Hire good consultants.
    2) Assign senior staff in the architect's firm to coordinate the consultants' plans.
    3) Make sure the consultants talk to each other and coordinate among themselves.
    4) Have frequent coordination meetings with all the consultants present.

24. If the county wants you to design a enjoyable, comfortable, and lovable library for the people, what is the best way to achieve this goal?
    1) Conduct a telephone survey of people in the county.
    2) Conduct a paper survey of the likely users of the library.
    3) Put an advertisement in a local newspaper.
    4) Hold a meeting with people in the neighborhood.

25. What is the major structural consideration for the existing building?
    1) whether the existing building's bearing capacity is adequate for the dead load of the bookshelves
    2) whether the existing building meets the seismic code requirements
    3) whether the existing building can retain its historical features
    4) whether the existing building's bearing capacity is adequate for the dead load of the bookshelves and whether the existing building meets the seismic code requirements

26. Which of the following is the most cost effective and safest structural material to deal with the seismic threat for this project?
    1) wood
    2) steel
    3) concrete
    4) reinforced concrete

27. Which of the following has the least impact on this project?
    1) number of bathrooms needed
    2) number of handicap parking spaces needed
    3) number of total parking spaces needed
    4) landscape coverage ratio

28. Which of the following is unlikely to be used for this project?
    1) gypsum board ceiling in the general reading area
    2) insulation in interior partitions
    3) acoustical ceiling at the reference desk
    4) hardwood floor in the children's reading area

29. During plan check, the building official instructs the architect to add an equipment screen wall for the rooftop units (RTU) for HVAC equipment only if they are visible from two adjacent major streets by the building inspector after the RTU is installed. What is the best way for the architect to handle this in the bidding process?
    1) Ask the contractors to include the equipment screen wall as an allowance in the bid.
    2) Ask the contractors to include the equipment screen wall as an alternate bid.
    3) Ask the contractors to include the equipment screen wall as a change order.
    4) Ask the contractors to include the equipment screen wall as an order for a minor change in the work.

30. The civil and plumbing site and floor plans only show the water line up to 5' outside of the building, and does not show the connecting water line from water main to within 5' of the new addition portion of the building. There is a general note on the cover sheet of the plans instructing the contractors to construct a complete water supply system, including any connection to the water main from the new addition. The contractor did not give the complete set of the plans to the plumbing subcontractor during the bidding, and the plumbing subcontractor has not included the cost for the connecting water line from water main to within 5' of the building in his bid based on the civil and plumbing site and floor plans. Who should pay for the extra costs?
    1) The general contractor since a general note on the cover sheet covers this item.
    2) The owner since the plumbing subcontractor has not included the cost for this item in the original bid.
    3) The civil engineer or the plumbing engineer since they have not properly shown this item on the civil and plumbing site and floor plans.
    4) The architect or her insurance company since she is responsible for coordinating the civil engineer or the plumbing engineer's work, and should have caught this conflict in the plans.

31. After the contract is awarded, the contractor notices that he has the site plans, but does not have the legal description of the site and the owner's interest in the site. The contractor requests the legal description from the owner. Which of the following is correct?
    1) The owner can refuse to provide the information since it is private.
    2) The contractor shall go to the county recorder's office and obtain the information since the owner is not obliged to provide it.
    3) The owner shall provide the information to the contractor.
    4) The contractor shall request the information from the civil engineer, not the owner.

32. At a field visit, the architect notices a subcontractor standing on a foam cornice next to the parapet. She is concerned about the safety procedures at the site. What should she do?
    1) Instruct the subcontractor not to stand on the foam cornice since it may not be strong enough to support a human weight.
    2) Do nothing, since jobsite safety procedures are the contractor's sole responsibility.
    3) Notify the owner about her concerns.
    4) Notify the superintendent about her concerns.

33. Which of the following statements regarding the initial decision maker (IDM) and mediation is NOT true per A201–2007, General Conditions of the Contract for Construction?
    1) Only the architect can be the IDM.
    2) The owner and the contractor can choose someone other than the architect to be the IDM.
    3) For most claims, a decision by the IDM is a prerequisite for mediation.
    4) none of the above

34. Once the owner files a Notice of Completion, the contractor must file a lien within how many days?
    1) 30
    2) 60
    3) 90
    4) 120

35. Seven years after the project completion the county finds out the building has uneven settlement. Investigation reveals that the settlement was caused by defective foundation installation. Which of the following is true?
    1) The contractor is not liable because it is discovered after the one-year statute of limitations.
    2) The contractor is not liable because it is discovered after the two-year statute of limitations.
    3) The contractor is liable because it is discovered within the ten-year statute of limitations.
    4) The contractor is liable, because there are no statues of limitations for defective work.

36. Who should pay for a special inspection?
    1) the county
    2) the contractor
    3) the architect
    4) none of the above

37. Per A201–2007, General Conditions of the Contract for Construction, what is the minimum number of inspections that an architect needs to perform?
    1) two
    2) three
    3) four
    4) five

38. Which of the following does not affect life-safety in a library?
    1) adequate ventilation
    2) proper anchorage of structural items
    3) proper anchorage of non-structural items
    4) none of above

39. Which of the following is not a prerequisite for final payment?
    1) a certificate evidencing required insurances
    2) an affidavit showing that payrolls, bills for materials and equipment, etc. related to the project have been paid
    3) consent of surety if any
    4) notice of completion

40. Which of the following is not a "saturated zone"?
    1) areas below the water table
    2) a surface that cannot absorb more water
    3) subsurface areas between the high water table level and low water table level
    4) watershed of ground water

41. What does the substantial completion trigger?
    1) date of mechanic's lien
    2) the owner taking over responsibility for insurance
    3) date of final completion
    4) date of final inspection

42. Once the owner files a Notice of Cessation, a subcontractor must file a Claim of Lien in the Office of the County Recorder within how many days?
    1) 30
    2) 60
    3) 90
    4) 120

B. Mock Exam: General Section

***Note:** Each question has only one correct answer.*

43. The California Architects Practice Act (APA) and Corporation Codes require a Limited Liability Partnership with five or fewer licensed architects to maintain a liability insurance limit of not less than
    1) $250,000.00
    2) $500,000.00
    3) $1,000,000.00
    4) $2,000,000.00

44. The APA requires an architect to use a written contract including all of the following elements except:
    1) a description of the services to be provided by the architect
    2) a description of any basis of compensation applicable to the contract and method of payment agreed upon by both parties
    3) a description of the professional liability insurance limit
    4) the name and address of the architect and the name and address of the client

45. The APA and Corporation Codes require a corporation owned by an architect with five or fewer licensed architects to maintain a liability insurance limit of not less than
    1) $500,000.00
    2) $1,000,000.00
    3) $2,000,000.00
    4) none of above

46. According to APA, which of the following acts by an unlicensed person is considered practicing without a license or holding oneself out as architect?
    1) advertising on Craigslist that you are not an architect, but that you can do the same work as an architect: designs and plans for a remodel, room addition, or custom home
    2) telling your wife you are an architect
    3) distributing business cards bearing the words "registered building designer"
    4) stating on your resume that you have over 20 years of architectural experience

47. Which of the following statements is a condition where the architect is NOT required to stamp and sign the plans?
    1) The architect prepares plans to the city for the first time.
    2) The architect prepares plans for public buildings owned by a county.
    3) The architect works as a principal in charge of the project.
    4) The architect exercises responsible control on a project as an employee in an architectural firm owned by another architect.

48. Which of the following is a situation when the architect is exempt from negligent inspection?
    1) The architect provides inspection services for his mother-in-law's house that he designed.
    2) The architect provides structural inspection services at the scene of a declared state emergency after a major earthquake at the request of a public official on a pro bono basis within 30 days of the declared emergency.
    3) The architect provides structural inspection services at the scene of a declared state emergency after a major earthquake at the request of a public official on a pro bono basis within 90 days of the declared emergency.
    4) none of the above

49. Which of the following is a situation when the architect is required to have a written contract with the client?
    1) The architect provides professional services to design a restaurant for a client he has been working with for many years on similar projects.
    2) The architect provides professional services on a pro bono basis.
    3) The architect provides professional services to a registered professional engineer.
    4) The architect provides a client professional services for preliminary studies of a project.

50. Which of the following statements is true?
    1) A registered civil engineer can stamp and sign all plans for a two-story single-family home.
    2) If an architect's license has expired for six years, he just needs to file a form prescribed by the board and pay the delinquency fee to renew his license again.
    3) An architect cannot form a partnership with an unlicensed person.
    4) An unlicensed person is not allowed to prepare plans for a two-story custom home of wood-frame construction.

51. Which of the following statements regarding CEQA is true?
    1) Once a lead agency decides to prepare a Negative Declaration, no public review is required.
    2) Once a lead agency decides to prepare an EIR, they will give public notice.
    3) After reviewing the final EIR and making a decision on the project, a local agency will file a Notice of Determination with the County Clerk.
    4) After reviewing the final EIR and making a decision on the project, a local agency will file a Notice of Determination with the Office of Planning & Research.

52. Which of the following statements is not true?
    1) The State Clearinghouse (SCH) also functions as the "State Single Point of Contact" for coordinating state and local review of applications for federal grants or loans under select state programs (Presidential Executive Order 12372).
    2) The Executive Order does not replace public participation, comment, or review requirements of other federal laws, such as the National Environmental Quality Act (NEQA), but gives the states an additional mechanism to ensure federal agency responsiveness to state and local concerns.
    3) OPR makes information about notices of exemption, notices of preparation, notices of determination, and notices of completion available to the public through CEQAnet (CEQA Guidelines Section 15023 (h)).
    4) The SCH coordinates the State review of the CEQA environmental documents.

53. Per the Essential Services Buildings Seismic Safety Act, which of the following statements is not true?
    1) The Essential Services Buildings Seismic Safety Act was passed by the California Legislature in 1986.
    2) Per the Essential Services Buildings Seismic Safety Act, the nonstructural components vital to the operation of essential services buildings shall also be able to resist, insofar as practical, the forces generated by earthquakes, gravity, fire, and winds.
    3) "Essential services building" means any building (or portion of building), including buildings designed and constructed for public agencies that are used, or designed to be used, as a fire station, hospital, police station, emergency operations center, California Highway Patrol office, sheriff's office, or emergency communication dispatch center.
    4) Responsibility for enforcement of the Essential Services Buildings Seismic Safety Act falls to the local building jurisdiction for locally owned or leased facilities.

54. Which of the following is exempt from the Essential Services Buildings Seismic Safety Act?
    1) essential services buildings of one-story Type V and Type II N construction that are 2,000 square feet or less in floor area
    2) essential services buildings of two-story Type V and Type II N construction that are 2,000 square feet or less in floor area
    3) essential services buildings of one-story Type V and Type II N construction that are 5,000 square feet or less in floor area
    4) essential services buildings of two-story Type V and Type II N construction that are 5,000 square feet or less in floor area

55. In addition to the California Energy Code, which of the following codes also has energy conservation provisions?
    1) California Plumbing Code
    2) California Elevator Safety Construction Code
    3) California Historical Building Code
    4) California Fire Code

56. How many parts does the California Building Standards Code, or Title 24 of the California Code of Regulations (CCR) have?
    1) 8
    2) 12
    3) 16
    4) 24

57. The Seismic Design Category (SDC) of a structure is NOT determined by its
    1) design spectral response acceleration coefficients
    2) occupancy category
    3) height
    4) footprint

58. Which of the following regarding the Seismic Design Category (SDC) of a structure is true?
    1) Buildings in SDC A need to meet more restrictive earthquake requirements than buildings in SDC C.
    2) Buildings in SDC C need to meet more restrictive earthquake requirements than buildings in SDC A.
    3) SDC includes categories A through D.
    4) SDC includes categories A through G.

59. In California, suspended ceilings must have diagonal bracing wires within the attic space, connecting the metal grid to the structure above. One set of four brace wires should be spaced
    1) 12 feet on center each way
    2) 6 feet on center each way
    3) 12 feet on center
    4) 6 feet on center

60. Glass adjacent to or within entrance or exit doors must be
    1) safety glass (i.e. tempered or laminated)
    2) safety glass (i.e. tempered or wired)
    3) safety glass (i.e. tempered or diagonal wired)
    4) safety glass (i.e. tempered or annealed)

61. Per the California Building Code (CBC), curb ramps shall have a detectable warning that
    1) extends the full width and depth of the curb ramp, excluding the flared sides, inside the grooved border
    2) extends the full width and depth of the curb ramp, including the flared sides, inside the grooved border
    3) extends the full width and two feet depth of the curb ramp, excluding the flared sides, inside the grooved border
    4) extends the full width and three feet depth of the curb ramp, excluding the flared sides, inside the grooved border.

62. Per the CBC, which of the following are acceptable dimensions for an accessible shower stall?
    1) 30" x 60"
    2) 36" x 60"
    3) 42" x 48"
    4) all of the above

63. Per the CBC, with a required occupant capacity of 100, the minimum corridor width shall not be less than
    1) 36"
    2) 44"
    3) 48"
    4) 60"

64. Per the CBC, which of the following is true?
    1) Ramps shall have a minimum clear width of 44 inches, unless required to be wider by some other provision of this code.
    2) Ramps serving entrances to buildings, where the ramp is the only exit discharge path and serves an occupant load of 500 or more, shall have a minimum clear width of 60 inches.
    3) Ramps serving Group R occupancies may be 36 inches clear width when the occupant load is 50 or less.
    4) all of the above

65. Which of the following is not subject to the California Energy Code (CEC)?
    1) buildings of Occupancy Group U
    2) buildings that are unconditioned
    3) low-rise residential buildings that are heated with a wood heater or another non-mechanical heating system
    4) none of the above

66. Per the CEC, only habitable floors that have at least ______ percent of their volume above grade as defined in the CBC shall be counted in determining how many habitable stories a building has.
    1) 40
    2) 50
    3) 60
    4) 80

67. In California, plans for a hospital building are submitted to the following governing agencies for a plan check:
    1) the city
    2) the health department
    3) the Office of Statewide Health Planning and Development
    4) all of the above

68. Per the California Health and Safety Code (CHSC), which of the following is required for a public pool?
 1) an anti-entrapment grate
 2) a backup safety system
 3) an approved safety pool cover
 4) all of the above
 5) none of the above

69. If a mechanics lien is filed against a property, the property owner
 1) is ultimately responsible for payment — even if she has already paid the prime contractor
 2) is no longer responsible for payment — as long as she has already paid the prime contractor
 3) is no longer responsible for payment — as long as she can show proof that she has already paid the prime contractor
 4) none of the above

70. A subcontractor or material supplier has __________ days after beginning work or delivering materials to serve the property owner a Preliminary Notice.
 1) 20
 2) 30
 3) 60
 4) 90

71. The California Coastal Zone generally extends seaward to the state's outer limit of jurisdiction, including all offshore islands, and extends inland generally _________ yards from the mean high tide line of the sea.
 1) 500
 2) 1,000
 3) 2,000
 4) 5,000

72. The California Clean Air Act was signed into law in
 1) 1972
 2) 1975
 3) 1988
 4) 1997

73. The Field Act requires a base shear calculation, and that school buildings must be able to withstand lateral forces equal to at least __________ of the building's total mass.
 1) 3%
 2) 5%
 3) 10%
 4) 20%

74. The California Senate SB 1953 sets a series of deadlines for compliance with the earlier Hospital Facilities Seismic Safety Act (HFSSA) of 1973. It requires that by the year 2008 _____________ must be able to withstand a major earthquake and all nonstructural mechanical, electrical, and plumbing systems in critical-care areas must be braced and anchored to assure survivability.
   1) all new general acute-care hospitals
   2) all new general acute-care inpatient buildings
   3) all general acute-care hospitals
   4) all general acute-care inpatient buildings

75. The "Public entity," as used in the Public Contract Code, means
   1) the state
   2) the city
   3) a political subdivision
   4) all of the above

76. Which of the following statements regarding the values of California's wetlands is not accurate?
   1) Biological Diversity: Wetlands provide important habitats for diverse communities of plants and animals, including about 30 percent of the federally listed threatened or endangered species.
   2) Waterfowl Habitat: Wetlands provide the principal habitat for migratory waterfowl. California provides critical wintering habitats for millions of waterfowl migrating along the Pacific Flyway, which extends from Canada to Mexico.
   3) Water Quality: Wetlands absorb and filter pollutants that could otherwise degrade ground water or the water quality of rivers, lakes, and estuaries.
   4) Flood Control: Wetlands detain flood flows, reducing the size and destructiveness of floods.

77. Coastal California, also known as the California Coastline and the Golden Coast, refers to the coastal regions of the U.S. state of California. The term is based primarily on the following attributes EXCEPT
   1) culture
   2) economy
   3) politics
   4) geography

78. Per Sections 2081(b) and (c) of the California Endangered Species Act, if a State listed threatened and endangered species is found on a project site, what action should the architect take?
   1) Assist the owner to obtain a variance.
   2) Assist the owner to obtain an incidental take permit.
   3) Assist the owner to prepare an EIR.
   4) Assist the owner to obtain a CUP.

79. Which of the following types of soil is the most expansive?
    1) clay
    2) silt
    3) gravel
    4) sandy

80. Which of the following can cause soil liquefaction?
    1) standing underground water
    2) an underground spring
    3) earthquakes
    4) all of the above

81. Which of the following is an effective way to prevent soil erosion flooding?
    1) silt fencing
    2) using a concrete patio around the building
    3) maintaining a 2% slope away from the building
    4) all of the above

82. What is the minimum ventilation rate for an office per ASHRAE?
    1) 5 cfm/person
    2) 10 cfm/person
    3) 30 cfm/person
    4) 50 cfm/person

83. What is the best way to control surface vapor pressure and related mold?
    1) humidifiers
    2) HEPA filters
    3) cooling the room
    4) ventilation

84. During programming, which of the following will an architect typically consider in an initial project cost estimate?
    1) cost of utilities
    2) labor cost
    3) materials cost
    4) inflation

85. Which of the following will trigger a special inspection?
    1) a field weld
    2) using concrete with a strength of 3,000 psi or more for better quality per the architect's request
    3) using concrete with a strength of 3,000 psi or more per the structural engineer's recommendation
    4) Both 1) and 3)

86. Which of the following is the *minimum* qualification to be an inspector for Essential Service Buildings?
    1) an inspector certified by DSA
    2) an inspector certified by ICBO
    3) an inspector certified by OSHPD
    4) an inspector certified by Essential Service Buildings Department of California

87. Per the energy code, a signed Installation Certificate should be
    1) posted at the jobsite
    2) provided to the building owner at occupancy
    3) Either 1) or 2)
    4) Both 1) and 2)

88. Per the energy code, who of the following should verify the refrigerant charge?
    1) a mechanical contractor
    2) an independent building inspector
    3) a HERS rater
    4) a general contractor

89. Variances are used to
    1) change design guidelines of a city
    2) change the local zoning guidelines of a city
    3) allow a property to be constructed in a manner basically consistent with the established zoning regulations with minor variations
    4) allow for a use that is completely different from the established zoning regulations

90. Which of the following is not true regarding a Conditional Use Permit (CUP)?
    1) A CUP usually requires a public hearing.
    2) A CUP is typically approved by a city or county.
    3) Any city or county zoning ordinance can list examples of common uses allowed with a conditional use permit.
    4) A CUP allows special use in a manner basically consistent with the established zoning regulations with minor variations.

91. The cost of making an existing building accessible is 30% of the overall renovation construction cost. Which of the following actions should an architect take?
    1) Advise the owner to make the building accessible.
    2) Advise the owner to request an unreasonable hardship waiver.
    3) Advise the owner about an unreasonable hardship waiver, and the potential lawsuits because of the non-compliance.
    4) Advise the owner about an unreasonable hardship waiver, and inform the owner that as long as the governing agencies approve the waiver, there is no potential for lawsuits for the non-compliance of laws.

92. Per ADA, the clear floor or ground space for a wheel chair is
    1) 30"x42"
    2) 30"x48"
    3) 36"x42"
    4) 36"x48"

93. Which of the following is not related to the National Environmental Policy Act (NEPA)?
    1) Environmental Impact Report (EIR)
    2) Categorical Exclusion (CE)
    3) Finding of No Significant Impact (FONSI)
    4) Council on Environmental Quality (CEQ)

94. Which of the following is not a term used in FAA determinations?
    1) no objection
    2) denied
    3) conditional approval
    4) ALP

95. Which of the following is actively engaged in ISO 14000 (environmental)?
    1) ANSI
    2) ASTM
    3) Factory Mutual
    4) UL

96. How many Regional Water Quality Control Boards (Regional Boards) does California have?
    1) four
    2) five
    3) nine
    4) twelve

97. For bathrooms and janitor rooms in retail buildings that process food or sell pre-packaged foods, environmental health departments will typically accept the following EXCEPT:
    1) a floor with VCT flooring and a 6" continuous vinyl base
    2) a ceramic tile floor with a 6" high continuous ceramic tile cove base
    3) a sealed concrete floor and a 6" high ceramic tile with a slim foot base
    4) a sheet vinyl floor with a 6" continuous wrap-up base

98. For a project that has impact on the nation's aquatic resources, which of the following agencies will review the application and grant a permit?
    1) Coastal Commission
    2) EPA
    3) FEMA
    4) U.S. Army Corps of Engineers

99. What should an architect do if there is a conflict between ADA and CBC?
    1) Follow the most restrictive requirement.
    2) Follow ADA since it is a federal law.
    3) Follow CBC since it is California Building Code and tailored to California.
    4) Let the building official decide which one to use.

100. Which of the following is a responsibility of a civil engineer in the schematic design stage?
    1) boundary survey and topography survey
    2) schematic parking lot layout
    3) schematic planting plan
    4) SWPPP

101. After setting a date for your presentation to the Design Review Board (DRB) for your project, what does the Planning Department do?
    1) notifies residents adjacent to your project site of the date for your presentation to DRB so that the residents can attend the presentation and comment on your project
    2) notifies all city departments to solicit their comments on your project
    3) notifies all other governing agencies to solicit their comments on your project
    4) forwards your project to the fire department for comments

102. Typically which of the following will contribute documents for the presentation to the Design Review Board (DRB)?
    1) electrical engineer
    2) mechanical engineer
    3) plumbing engineer
    4) landscape architect

103. Which of the following typically requires a separate submittal?
    1) building department
    2) environmental health department
    3) fire department
    4) planning department

104. A client wants an architect to guarantee the highest standard of architectural service in the contract. What should the architect do?
    1) Incorporate the client's request in the contract.
    2) Negotiate with the client for a higher fee to guarantee the highest standard of architectural service.
    3) Inform the architect's insurance company about the client's request.
    4) Inform the client that s/he can only provide an architectural service that is ordinarily provided by architects practicing under the same or similar circumstances.

105. An architect is doing an addition project for a home owner. The client wants the architect to provide professional liability insurance with a limit of two million dollars. The architect currently does not have professional liability insurance. What options does the architect have?
    1) Purchase the professional liability insurance per the client's request.
    2) Negotiate with the client and ask the client to pay for the premium of the professional liability insurance.
    3) Negotiate with the client and try to add a clause in the contract indicating that the architect's total professional liability shall not be more than the architectural fee for project.
    4) all of the above

106. CalGreen requires construction waste reduction of at least
    1) 25%
    2) 30%
    3) 50%
    4) 60%

107. Which of the following does CalGreen require for permit?
    1) ASHRAE calculations
    2) energy management plan
    3) waste management plan
    4) water conservation plan

108. Which of the following does CBC Chapter 17 require as a part of the permit submittal package?
    1) a soils report
    2) a schedule of special inspections
    3) structural calculations
    4) all of the above

109. Which of the following does not need to be braced in a seismic zone?
    1) a 2,000-lb air handler
    2) a 1,000-lb condensing unit on a vibration isolator
    3) a 200-lb furnace
    4) a 30-lb pump mounted five feet above the floor

110. Which of the following use galvanized steel trapeze brackets to anchor?
    1) conduits and pipes
    2) recessed ceiling lights
    3) bookshelves
    4) water heaters

111. How long shall public notices of a project be given in a newspaper of general circulation published in the county in which the project is located, or in a designated trade paper of general circulation?
   1) once a week for at least two consecutive weeks
   2) once a week for at least two consecutive weeks or once a week for more than two consecutive weeks if the longer period of advertising is deemed necessary by the department
   3) twice a week for at least two consecutive weeks
   4) twice a week for at least two consecutive weeks or once a week for more than two consecutive weeks if the longer period of advertising is deemed necessary by the department

112. At a job site visit, an architect notices that the roofing contractor is not following safety procedures. What should the architect do?
   1) Tell the roofing contractor to follow the safety procedures.
   2) Tell the roofing contractor to follow the safety procedures and also inform the general contractor.
   3) Inform the general contractor in writing.
   4) Inform the general contractor and the owner in writing.

113. If the construction is delayed by a fire at the job site, what should the architect do?
   1) Issue a change order to extend the contract time as determined by the architect to be reasonable.
   2) Inform the contractor to keep the original schedule because the contractor is responsible for fire prevention at the job site.
   3) Discuss with the owner, and inform the contractor about the owner's decision on the project schedule.
   4) Have a meeting with the owner and contractor, and have the owner and contractor decide on the project schedule.

114. During construction of a renovation project, the contractor discovers asbestos in the ceiling and attic. What should the contractor do?
   1) Stop the construction and inform the owner and architect in writing.
   2) Remove the asbestos and continue the construction at no extra cost.
   3) Remove the asbestos and send a change order request to the owner.
   4) Hire a specialist to investigate the situation at the contractor's cost.

115. What is the most likely sequence to resolve disputes between an owner and contractor?
   1) initial decision, mediation, arbitration, and litigation
   2) initial decision, arbitration, mediation, and litigation
   3) initial decision, litigation, mediation, and arbitration
   4) initial decision, mediation, litigation, and arbitration

116. During construction, minor changes do not affect contract time or contract sum of the project if issued as a(an)
   1) addendum
   2) change order
   3) construction change directive
   4) architect's supplemental instruction

117. How is the schedule of values determined?
   1) by the contractor, subject to the architect's review
   2) by the contractor, subject to the owner's review
   3) by the contractor and owner
   4) by the contractor, owner, and architect

118. An architect notices that the contractor's application for payment includes the materials stored at the site and materials stored off the site. What should the architect do?
   1) Approve payment for the materials stored at the site, and deny payment for the materials stored off the site.
   2) Approve payment for the materials stored at the site, and complete more research on payment for the materials stored off the site, since these materials might qualify for payment also.
   3) Approve payment for the materials stored at the site and payment for the materials stored off the site.
   4) Approve payment for the materials already installed.

119. Warranties required by the contract documents shall typically commerce on the date of
   1) substantial completion
   2) completion of the punch list items
   3) final completion
   4) passing the final inspection

120. Who of the following is required to file a preliminary notice to be able to record a lien?
   1) the architect
   2) the contractor
   3) a sub-contractor
   4) none of the above

Chapter Four

CSE Mock Exam Solutions

A. Mock Exam Answers and Explanations: Project Scenario Section

Note: If you answer 60% of the questions correctly, you will pass the exam.

1. Answer: 2) Based on the plans on previous pages, the style of the existing building is Spanish Style.

   The following are incorrect answers:
   - Colonial Style
   - Mediterranean Style
   - Arts and Crafts Style

   As an architect, you should know the basic architectural styles. If you do not know, you can probably guess the correct answer since California has a long history of mission buildings in Spanish Style.

2. Answer: 1) The building should be in occupancy group A (or A-3 to be more accurate) for the purpose of code analysis.

   The existing building is a custom home, but it will be converted into a library. You should use its intended function, as a library, to determine its occupancy group. An architect should be familiar with various occupancy groups, and know what kinds of buildings should be in which group.

   See California Building Code (CBC), Chapter 3, Use and Occupancy Classification, at the following link:
   http://publicecodes.cyberregs.com/st/ca/st/b200v10/st_ca_st_b200v10_3_sec001.htm

   - Occupancy Group A is Assembly Group.
   - Occupancy Group B is Business Group.
   - Occupancy Group E is Educational Group.
   - Occupancy Group R is Residential Group.

   You only need to know the frequently used sections of the CBC well (about 20% of the content for CBC). My other book, ***Building Construction:*** *Project Management, Construction Administration, Drawings, Specs, Detailing Tips, Schedules, Checklists, and Secrets Others Don't Tell You* (ISBN-13: 978-0984374144) has detailed discussions on this subject.

3. Answer: 1) R-11 is unlikely to be used in the exterior walls or roof of the new construction.

   Because of the Title-24 requirements, the insulation for exterior walls is typically R-13 or R-19, and the insulation for roofs is typically R-19 or R-30. R-11 is sometimes used for interior partitions.

4. Answer: 4) Based on CBC, Table 503 Allowable Building Heights and Areas, the most cost effective type of construction an architect should choose for this building for the purpose of code analysis is Type V-B.

   Per Table 503, the *basic* allowable area is 6,000 s.f. for Type V-B construction.

   Per CBC, Section 506.3, Automatic sprinkler system increase: "Where a building is equipped throughout with an approved automatic sprinkler system in accordance with Section 903.3.1.1, the building area limitation in Table 503 is permitted to be increased by an additional 200 percent (Is = 2) for buildings with more than one story above grade plane and an additional 300 percent (Is = 3) for buildings with no more than one story above grade plane."

   Per the Project Scope, the building will be fully-sprinklered.

   The *increased* allowable area for this building is 2 x 6,000 s.f. = 12, 000 s.f. > 11,800 s.f (the total square footage of both the existing and new addition of the building).

5. Answer: 4) Arranging the spaces to provide equal access on the first floor is the most cost effective way to comply with accessibility requirements.

   Adding an elevator or a wheel chair lift will cost extra. Providing a refuge area will not provide equal access.

6. Answer: 1) You are working on the project's detailed program. You will be unlikely to solicit information on spatial requirements from library readers, because they know very little on this subject.

   The Director of Orange County Public Library (OCPL) and Director of County Facilities Management are good sources for this information.

7. Answer: 3) If there are conflicts between various county departments regarding the spatial requirements, you should follow the instructions of the Director of County Facilities Management for this project because he is the point of contact for the county.

The following are incorrect answers:

- library readers (They are not part of county departments.)
- Director of Orange County Public Library (OCPL) (He is not the point of contact for the county.)
- county supervisors (They are not the point of contact for the county.)

8. Answer: 4) All the following may affect your project.
   - The existing building is a custom home, and may be subject to the Conditions, Covenants, and Restrictions (CC&R) of the homeowners association (**HOA**).
   - The Environmental Impact Report (**EIR**) definitely affects the project.
   - **The fire department** has very specific and detailed requirements as part of the municipal codes.

9. Answer: 2) The children's reading area requires proper separation from other areas because children tend to make noise and can affect other readers.

10. Answer: 3) The general reading area should not have priority for the first floor spaces.

    The following should have priority for the first floor space:
    - The **front desk** needs to be on the first floor and near the entrance.
    - The **reference desk** should be on the first floor and visible from the front desk.
    - The **bookshelves and isolated reading area** should be on the first floor because bookshelves are very heavy, and you should place as many as possible here.

11. Answer: 4) In addition to schedule and budget, zoning regulations are the most important factor determining the feasibility of this project.

    The existing building is a custom-home, and zoning regulations may affect the feasibility of converting the building into a library.

    The following should not be an issue in Newport Beach and are incorrect answers:
    - availability of labors
    - demography
    - infrastructure

12. Answer: 1) A conditional use permit (CUP) application is probably necessary for this project.

    Variances are used to allow a property to be constructed in a manner *basically consistent with* the established zoning regulations with *minor* variations. Converting a custom home to a library is not a *minor* variation.

    A **conditional use permit (CUP)** allows a local agency to consider special uses through a public hearing process. Such special uses are usually essential or *desirable* to a particular community, but *not allowed by zoning ordinances*. This is applicable to our project.

Per Wikipedia:
Eminent domain is "the power to *take private property for public use* by a state or national government." Eminent domain does not apply to our project.

Nonconforming use is "a type of zoning variance where a parcel of land may be given an exception from current zoning ordinances due to improvements made by a prior owner or before the current zoning ordinances made the desired use non-conforming under local law." We are not using the custom home as a residence, we are converting it to a library. Nonconforming use does not apply to our project.

13. Answer: 4) none of the above
UL's five businesses include: product safety, verification services, life & health, knowledge services, and environment. These demonstrate their "expanding breadth of expertise and growing range of services to offer solutions needed in a constantly evolving world."

See the following link:
http://www.ul.com/global/eng/pages/aboutul/whatwedo/

14. Answer: 4) all of the above

Per the California Coastal Act, Section 30233, diking, filling, dredging, and continued movements of sediment and nutrients can all be permitted on wetlands under certain conditions.

15. Answer: 1) Per the California Coastal Act, the City of Newport Beach Design Review Board has the designated authority to approve this project.

The designated authority should fall on a local agency/entity made up of representatives from the local community.

The following are incorrect answers:
- City of Newport Beach Planning Department (This department is not made up of representatives from the local community,)
- Californian Coastal Commission (The commission is not a local agency.)
- Orange County Board of Supervisors (This group is not made up of representatives from the local community.)

16. Answer: 1) The Department of Fish and Game is a state agency that has jurisdiction over this project because the project is next to Back Bay.

The following are incorrect answers:
- Division of the State Architect (DSA) (They have no jurisdiction over this project.)
- Office of Statewide Health Planning and Development (They have no jurisdiction over this project.)
- Water Quality Control Board (They are not a state agency.)

17. Answer: 4) The planning department wants to have minimum lighting levels to avoid light pollution. The police department wants to have bright lighting levels to avoid crimes. As an architect, you should work with both the planning department and the police department to come up with a compromise or a solution.

    The following are incorrect answers:
    - Follow the planning department's instructions to have minimum lighting levels since the planning department is in charge of the overall review of the project. (This will not satisfy the police department's requirements.)
    - Follow the police department's instructions to have bright lighting levels since avoiding crimes is a top priority. (This will not satisfy the planning department's requirements.)
    - Ask the city's building official for a final decision. (The city's building official is not the person who has the final say on this issue.)

    Architecture is a balancing act, and an architect needs to know how to deal with the conflicting demands of different parties.

18. Answer: 3) A surveyor can prepare a CLTA map with both topography and boundary information.

    CLTA stands for the **California Land Title Association**, which is a nonprofit corporation founded in 1907.

    CLTA's official website is:
    http://www.clta.org/

    A civil engineer can prepare a survey map with topography, but is NOT authorized to prepare a survey map with boundary information.

    A landscape architect typically does not prepare any kind of survey map.

19. Answer: 3) The county has hired a civil engineer directly. The civil engineer has missed the project deadline several times, and delayed the architect's project schedule. The architect should contact the county and let the county deal with the delay.

    Since the county has hired a civil engineer directly, all correspondence should go through the county.

    The following are incorrect answers:
    - Contact the civil engineer directly. (Never do this.)
    - Contact the civil engineer and copy the county in all correspondence. (You can ONLY do this if the county specifically authorizes you to do so.)

20. Answer: 2) One of the mechanical ducts is too large and cannot fit between the ceiling and the structural members above. The most cost effective solution would be to have the mechanical engineer revise the mechanical plans.

    The following are incorrect answers:
    - Have the structural engineer revise the structural plans. (This is too expensive.)
    - Have the structural engineer and mechanical engineer communicate with each other in order to resolve this issue. (You are the architect, and you should tell them what to do in regards to this issue.)
    - Ask a contractor for suggestions. (A contractor has nothing to do with this.)

21. Answer: 1) Holding a design charrette is an effective way to obtain stakeholder input.

    The following are incorrect answers:
    - Schedule a scope meeting with all departments of the city. (The city is only one of the stakeholders, and a scope meeting is typically held with the client to find out about the scope of the project.)
    - Have a dinner party with all the stakeholders. (You may not get any input at this sort of event.)
    - Put an advertisement on local TV. (You may not get any input through this method.)

22. Answer: 3) Keeping a checklist is an effective way to ensure quality assurance in an architect's office.

    The following are incorrect answers:
    - Hold a design charrette. (This is good for obtaining input from stakeholders, but not good for quality assurance.)
    - Have a manufacturer's representative give a lunch seminar on a product used in the project. (This is only helpful for one product, not for all the other products specified in the set of documents.)

23. Answer: 1) The *most* effective way to ensure quality assurance for consultants' plans is to hire good consultants to begin with.

    All the following will help ensure quality assurance for consultants' plans, but they are NOT the *most* effective way:
    - Assign senior staff in the architect's firm to coordinate the consultants' plans.
    - Make sure the consultants talk to each other and coordinate among themselves.
    - Have frequent coordination meetings with all the consultants present.

24. Answer: 2) If the county wants you to design a enjoyable, comfortable, and lovable library for the people, the best way to achieve this goal is to conduct a paper survey of the likely users of the library.

All the following will help but they are NOT the best way:

- Conduct a telephone survey of people in the county.
- Put an advertisement in a local newspaper.
- Hold a meeting with people in the neighborhood.

25. Answer: 4) The major structural consideration for the existing building is whether the existing building's bearing capacity is adequate for the dead load of the bookshelves and whether the existing building meets the seismic code requirements.

    We put the partially correct answers at the beginning, and the correct answer at the end. This is to train you to be patient enough to read through all four answers before you respond.

    The following are incorrect answers:
    - whether the existing building's bearing capacity is adequate for the dead load of the bookshelves (This is correct, but not the best answer.)
    - whether the existing building meets the seismic code requirements (This is correct, but not the best answer.)
    - whether the existing building can retain its historical features (This is not a structural consideration.)

26. Answer: 1) The most cost effective and safest structural material to deal with the seismic threat for this project is wood. Wood is cost effective, safe, and gives a lot of warnings before it fails.

    The following are incorrect answers:
    - steel (This material is not cost effective.)
    - concrete (This material is not safe, and gives few warnings before it fails.)
    - reinforced concrete (This material is not cost effective.)

27. Answer: 4) Landscape coverage ratio has the least impact on this project because the site area is two acres, and the average landscape coverage ratio is 10% to 15%. We should have plenty of landscape areas to meet the ratio.

    The following are incorrect answers:
    - number of bathrooms needed (This number will affect the building construction cost.)
    - number of handicap parking spaces needed (Handicap parking spaces are more expensive to build than standard parking spaces.)
    - number of total parking spaces needed (Parking area is more expensive to build than landscape area.)

28. Answer: 1) Gypsum board ceiling in the general reading area is unlikely to be used for this project because gypsum board ceilings are not good at absorbing sound, and cannot keep the noise level down.

    The following are incorrect answers:
    - insulation in interior partitions (This will probably be needed to keep the noise level down.)
    - acoustical ceiling at the reference desk (This will probably be needed to keep the noise level down.)
    - hardwood floor in children's reading area (Hardwood floor is easy to clean and good for kids to crawl around on. Therefore it’s a good choice for the children's reading area.)

29. Answer: 1)The best way to handle this in the bidding process is to ask the contractors to include the equipment screen wall as an allowance in the bid.

    See A201–2007, General Conditions of the Contract for Construction, Section 3.8. A change order and an order for a minor change in the Work are used during construction, not the bidding process.

30. Answer: 1) The contractor will pay for the missed item since a general note on the cover sheet covers this item.

    Per A201–2007, General Conditions of the Contract for Construction, the contract documents are complementary, and what is required by one shall be as binding as if required by all. The contractor should have coordinated with all his subcontractors and submitted the bid based on the entire set of contract documents, not just one portion.

    On the other hand, as an architect, you should also coordinate with your consultants to avoid conflicts within the contract documents. This shows you the importance of coordinating the plans and specifications.

31. Answer: 3) The owner shall provide the information to the contractor within 15 days of receiving the request per A201–2007, General Conditions of the Contract for Construction.

32. Answer: 4)
    Jobsite safety procedures are the contractor’s sole responsibilities, but the architect should still notify the superintendent about her concerns. She should never instruct a subcontractor what to do. Notifying the owner may be helpful, but it is not the best answer.

33. Answer: 1) The following statement regarding the initial decision maker (IDM) and mediation is NOT true per A201–2007, General Conditions of the Contract for Construction:
    - Only the architect can be the IDM.

See A201–2007 commentary, General Conditions of the Contract for Construction, Introduction.

Pay attention to the word "NOT." We are looking for the incorrect statement.

The following are correct statements but incorrect answers:

- The owner and the contractor can choose someone other than the architect to be the IDM.
- For most claims, a decision by the IDM is a prerequisite for mediation.

34. Answer: 2) Once the owner files a **Notice of Completion**, the contractor must file a lien within 60 days.

    See the following link:
    http://www.cab.ca.gov/pdf/misc/mechanics_lien_law.pdf

35. Answer: 3) The contractor is liable because it is discovered within the ten-year statute of limitations.

    All licensed contractors have to provide a one-year warranty for new construction and remodel projects, a four-year warranty for defective items installed under his license, and a ten-year warranty for structural defects.

    See the following link:
    http://www.professionalhomeinspection.net/pdf/contractor_law_1-4-10_yr.pdf

36. Answer: 4) none of the above

    Special inspections are paid by the owner, not the contractor. A201–2007 section 3.7.1 notes inspections shall be paid by the contractor, but does not mention "special inspections," which are different from standard inspections. CBC 1704 and 1705 define what special inspections are and clearly state that the owner pays for them.

37. Answer: 1) Per A201–2007, General Conditions of the Contract for Construction, the *minimum* number of inspections that an architect needs to perform is two. One inspection at punch walk, and the other before final payment.

    See A201–2007 commentary, General Conditions of the Contract for Construction, Section 9.8.3.

38. Answer: 4) none of above

Pay attention to the word "not."

All of the following affect life-safety in a library and are therefore the incorrect answers:
- adequate ventilation
- proper anchorage of structural items
- proper anchorage of non-structural items (This is critical for life-safety. Many people get hurt because of non-structural items such as bookshelves falling down during an earthquake.)

39. Answer: 4) Notice of completion is not a prerequisite for final payment.

Pay attention to the word "not."
All of the following are prerequisites for final payment and therefore not the correct answers:
- a certificate evidencing required insurances
- an affidavit showing that payrolls, bills for materials and equipment, etc. related to the project have been paid
- consent of surety if any

40. Answer: 3) subsurface areas between the high water table level and low water table level

Pay attention to the word "not."

A ***saturated zone*** is a surface that cannot absorb more water. This occurs after a heavy rainfall when soil can no longer absorb more water due to filling of all open spaces in the soil."

See the link:
http://www.ask.com/question/zone-of-saturation

"The Earth's crust can be divided into two regions: the ***saturated zone*** or ***phreatic zone*** (e.g., aquifers, aquitards, etc.), where all available spaces are filled with water, and the ***unsaturated zone*** (also called the vadose zone), where there are still pockets of air that contain some water, but can be filled with more water."

See the link:
http://en.wikipedia.org/wiki/Aquifer

The following are saturated zones and therefore not the correct answers:
- areas below the water table
- a surface that cannot absorb more water
- watershed of ground water

41. Answer: 2) The substantial completion triggers the owner taking over responsibility for insurance.

    The following are incorrect answers:
    - date of mechanic's lien (Final completion determines the date of the mechanic's lien.)
    - date of final completion
    - date of final inspection

42. Answer: 1) Once the owner files a **Notice of Cessation**, the subcontractor must file a Claim of Lien in the Office of the County Recorder within 30 days.

    Note:** The labor must cease for 30 days before the owner in entitled to file a **Notice of Cessation.

    See the following link:
    http://www.cab.ca.gov/pdf/misc/mechanics_lien_law.pdf

B. Mock Exam Answers and Explanations: General Section

43. Answer: 3) $1,000,000.00
    The California Architects Practice Act (APA) and Corporation Codes require a Limited Liability Partnership with five or fewer licensed architects to maintain a liability insurance limit of not less than $1,000,000.00.

    See the APA, Corporation Codes, Section 16956, "Security for Claims Against Limited Liability Partnership; Requirements; Evidence of Compliance."

44. Answer: 3) a description of the professional liability insurance limit
    Pay attention to the word "except."
    The APA requires an architect to use a written contract including all of the following elements except a description of the professional liability insurance limit.

    An architect should use a written contract including all of the following:
    1) a description of the services to be provided by the architect
    2) a description of any basis of compensation applicable to the contract and method of payment agreed upon by both parties
    3) the name, address, and license number of the architect and the name and address of the client
    4) a description of the procedure that the architect and client will use to accommodate additional services
    5) a description of the procedure to be used by either party to terminate the contract

    See the California Architects Practice Act, Section 5536.22, "Written Contract."

45. Answer: 4) none of above
    The APA and Corporation Codes do NOT require a corporation owned by an architect with five or fewer licensed architects to maintain a specific liability insurance limit.

    See APA, Article 7, and Moscone-Knox Professional Corporation Act at the following link:
    http://www.leginfo.ca.gov/cgi-bin/displaycode?section=corp&group=13001-14000&file=13400-13410

46. Answer: 3) distributing business cards bearing the words "registered building designer"

    See APA, Article 3, Section 5536.

    According to APA, the following acts by an unlicensed person are NOT considered practicing without a license or holding oneself out as architect:
    - advertising on Craigslist that you are not an architect, but that you can do the same work as an architect: designs and plans for a remodel, room addition, or custom home (This is a very confusing act, and per my e-mail to the CAB, this is NOT considered a violation because this person specifically stated s/he is NOT an architect.)

- telling your wife you are an architect (Your wife is not considered public, so you are not telling the public you are an architect.)
- stating on your resume that you have over 20 years of architectural experience (In this instance, you are looking for a job in an architectural firm, NOT practicing architecture yourself.)

47. Answer: 4) The architect exercises responsible control on a project as an employee in an architectural firm owned by another architect.

    An architect working as an employee is exempt from stamping and signing the plans. The licensed principal of the firm is supposed to stamp and sign the plans. See APA, Article 3, Section 5536.1.

    The following are conditions where the architect is required to stamp and sign the plans:
    - The architect prepares plans to the city for the first time. (Per APA, you are supposed to stamp and sign the plans even for the first city submittal.)
    - The architect prepares plans for public buildings owned by a county.
    - The architect works as a principal in charge of the project.

48. Answer: 2)
    The following is a situation when the architect is exempt from negligent inspection:
    - The architect provides structural inspection services at the scene of a declared state emergency after a major earthquake at the request of a public official on a pro bono basis within 30 days of the declared emergency.

    See APA, Article 3, Section 5536.27.

    The following are incorrect answers:
    - The architect provides inspection services for his mother-in-law's house that he designed.
    - The architect provides structural inspection services at the scene of a declared state emergency after a major earthquake at the request of a public official on a pro bono basis within 90 days of the declared emergency.
    - none of the above

49. Answer: 4)
    The following is a situation when the architect is required to have a written contract with the client:
    - The architect provides a client professional services for preliminary studies of a project.

    See APA, Article 3, Section 5536.22.

The following are situations when the architect is NOT required to have a written contract with the client:

- The architect provides professional services to design a restaurant for a client he has been working with for many years on similar projects.
- The architect provides professional services on a pro bono basis.
- The architect provides professional services to a registered professional engineer.

50. Answer: 1) A registered civil engineer can stamp and sign all plans for a two-story single-family home. (See APA, Article 3, Section 5537.1, "Exemptions; Structural Engineer.")

    The following are incorrect answers:

    - If an architect's license has expired for six years, he just needs to file a form prescribed by the board and pay the delinquency fee to renew his license again. (He may need to take the ARE exams again since his license has been expired more than five years. See APA, Article 6, Section 5600.3.)
    - An architect cannot form a partnership with an unlicensed person. (See APA, Article 3, Section 5535.2, "Partnerships with Non-Architects.")
    - An unlicensed person is not allowed to prepare plans for a two-story custom home of wood-frame construction. (See APA, Article 3, Section 5537, "Exemptions; Dwellings, Garages, Agricultural and Ranch Buildings; Supervision of Licensed Architect or Registered Engineer Required.")

51. Answer: 3) The following statement regarding CEQA is true:

    - After reviewing the final EIR and making a decision on the project, a *local* agency will file a Notice of Determination with the County Clerk.

    CEQA means the California Environmental Quality Act. EIR stands for an Environmental Impact Report.

    See the CEQA flow chart at the following links:
    http://ceres.ca.gov/ceqa/flowchart/
    http://www.opr.ca.gov/m_ceqa.php

    The following are incorrect answers:

    - Once a lead agency decides to prepare a Negative Declaration, no public review is required. (A public review is still required.)
    - Once a lead agency decides to prepare an EIR, it will give public notice. (Once a lead agency finishes drafting an EIR, it will give public notice.)
    - After reviewing the final EIR and making a decision on the project, a local agency will file a Notice of Determination with the Office of Planning & Research. (After reviewing the final EIR and making a decision on the project, a *state* agency will file a Notice of Determination with the Office of Planning & Research.)

52. Answer: 2) The following statement is not true:
    - The Executive Order does not replace public participation, comment, or review requirements of other federal laws, such as the National Environmental Quality Act (NEQA), but gives the states an additional mechanism to ensure federal agency responsiveness to state and local concerns.

    There is no such thing as the National Environmental *Quality* Act (NEQA). The correct term is the National Environmental *Policy* Act (NEPA).

    OPR stands for Office of Planning & Research.

    See information on State Clearinghouse at the following link:
    http://opr.ca.gov/m_stateclearinghouse.php

    Pay attention to the word "not." We are looking for the incorrect statement.

    The following are true statements and therefore the incorrect answers:
    - The State Clearinghouse (SCH) also functions as the "State Single Point of Contact" for coordinating state and local review of applications for federal grants or loans under select state programs (Presidential Executive Order 12372).
    - OPR makes information about notices of exemption, notices of preparation, notices of determination, and notices of completion available to the public through CEQAnet (CEQA Guidelines Section 15023 (h)).
    - The SCH coordinates the State review of the CEQA environmental documents.

53. Answer: 3) Per the Essential Services Buildings Seismic Safety Act, the following statement is not true:

    - "Essential services building" means any building (or portion of building), including buildings designed and constructed for public agencies that are used, or designed to be used, as a fire station, hospital, police station, emergency operations center, California Highway Patrol office, sheriff's office, or emergency communication dispatch center.

    A hospital is NOT an "essential services building." See "essential services building" definition at the following link:
    http://codes.lp.findlaw.com/cacode/HSC/1/d12.5/2/2/s16007

    Pay attention to the word "not." We are looking for the incorrect statement.

    The following are true statements and therefore the incorrect answers:
    - The Essential Services Buildings Seismic Safety Act was passed by the California Legislature in 1986.
    - Per the Essential Services Buildings Seismic Safety Act, the nonstructural components vital to the operation of essential services buildings shall also be able to

resist, insofar as practical, the forces generated by earthquakes, gravity, fire, and winds.

- Responsibility for enforcement of the Essential Services Buildings Seismic Safety Act falls to the local building jurisdiction for locally owned or leased facilities.

See the following link:
http://codes.lp.findlaw.com/cacode/HSC/1/d12.5/2

54. Answer: 1) The following is exempt from the Essential Services Buildings Seismic Safety Act:

- essential services buildings of one-story Type V and Type II N construction that are 2,000 square feet or less in floor area

The following are incorrect answers:

- essential services buildings of two-story Type V and Type II N construction that are 2,000 square feet or less in floor area
- essential services buildings of one-story Type V and Type II N construction that are 5,000 square feet or less in floor area
- essential services buildings of two-story Type V and Type II N construction that are 5,000 square feet or less in floor area

See the following link:
http://codes.lp.findlaw.com/cacode/HSC/1/d12.5/2/3/s16010

55. Answer: 3)
In addition to the California Energy Code, the following code also has energy conservation provisions:

- California Historical Building Code

California Historical Building Code, Chapter 8, Section 901.5 is titled "Energy Conservation." See the following link:
http://publicecodes.cyberregs.com/st/ca/st/b2200v07/st_ca_st_b2200v07_8-9_par005.htm

The following are incorrect answers:

- California Plumbing Code
- California Elevator Safety Construction Code
- California Fire Code

See the following links for the California Building Standards Code, or Title 24 of the California Code of Regulations (CCR):
http://www.bsc.ca.gov/pubs/codeson.aspx
http://www.dgs.ca.gov/dsa/Programs/progCodes/title24.aspx

56. Answer: 2) The California Building Standards Code, or Title 24 of the California Code of Regulations (CCR) has 12 parts. They are:
    - Part 1 - California Building Standards Administrative Code
    - Part 2 - California Building Code
    - Part 3 - California Electrical Code
    - Part 4 - California Mechanical Code
    - Part 5 - California Plumbing Code
    - Part 6 - California Energy Code
    - Part 7 - California Elevator Safety Construction Code
    - Part 8 - California Historical Building Code
    - Part 9 - California Fire Code
    - Part 10 - California Existing Building Code
    - Part 11 - California Green Building Standards Code
    - Part 12 - California Reference Standards Code

    See the following link:
    http://www.dgs.ca.gov/dsa/Programs/progCodes/title24.aspx

57. Answer: 4) The Seismic Design Category (SDC) of a structure is NOT determined by its footprint.

    The Seismic Design Category (SDC) of a structure is determined by its
    - design spectral response acceleration coefficients
    - occupancy category
    - height

    See CBC, 1613.5.6, Determination of seismic design category:
    "Occupancy Category I, II or III structures located where the mapped spectral response acceleration parameter at 1-second period, S_1, is greater than or equal to 0.75 shall be assigned to Seismic Design Category E. Occupancy Category IV structures located where the mapped spectral response acceleration parameter at 1-second period, S_1, is greater than or equal to 0.75 shall be assigned to Seismic Design Category F. All other structures shall be assigned to a seismic design category based on their **occupancy category** and the **design spectral response acceleration coefficients**, S_{DS} and S_{D1}, determined in accordance with Section 1613.5.4 or the site-specific procedures of ASCE 7. Each building and structure shall be assigned to the more severe seismic design category in accordance with Table 1613.5.6(1) or 1613.5.6(2), irrespective of the fundamental period of vibration of the structure, T."

    See the following link:
    http://publicecodes.cyberregs.com/st/ca/st/b200v10/st_ca_st_b200v10_16_par178.htm?bu2=undefined

    The structural **height** also affects the S_{DC}. For example, some low-rise structures are exempt from earthquake requirements. Detached one and two family wood-framed dwellings not more than two stories in height need to be designed using IRC requirements instead of IBC requirements.

58. Answer: 2) The following regarding the Seismic Design Category (SDC) of a structure is true:
    - Buildings in SDC C need to meet more restrictive earthquake requirements than buildings in SDC A.
    - SDC includes categories A through F.

    Occupancy group I are buildings with a *low hazard* to human life if they fail, such as agricultural facilities, minor storage facilities, and temporary facilities.

    Occupancy group II are other buildings *not* included in occupancy groups I, III, and IV.

    Occupancy group III are buildings with a *substantial hazard* to human life if they fail, such as public assemblies with an occupant load over 300; elementary or secondary schools, or daycare facilities with an occupant load over 250; colleges or adult education facilities with an occupant load over 500; jail and detention facilities; power-generating, potable water-treatment facilities, waste water-treatment facilities, or other public facilities not included in occupancy group IV; buildings with sufficient quantities of explosives or toxic substances and not included in occupancy group IV; or health care facilities with 50 or more resident patients, but without surgery or emergency treatment facilities.

    Occupancy group IV are buildings designated as *essential facilities*, such as hospitals or health care facilities with surgery or emergency treatment facilities; police, fire, and rescue stations or emergency vehicle garages; designated earthquake, hurricane, or other emergency shelters; designated communication, emergency preparedness, and operation centers for emergency responses; power-generating or other public facilities required as back up facilities for occupancy group IV structures; buildings with highly toxic substances exceeding the maximum allowable quantities of Table 307.1.(2); air traffic control centers, aviation control towers, or emergency aircraft hangers; structures with significant national defense functions; or water-treatment facilities required to maintain water pressure for fire suppression.

    Seismic Design Category A are buildings of **all** occupancy groups in areas expecting minor ground shaking. (Good soil)
    Seismic Design Category B are buildings of occupancy groups I, II, and III in areas expecting moderate ground shaking. (Stratified soil with good and poor soil)
    Seismic Design Category C are buildings of occupancy groups IV in areas expecting moderate ground shaking, and buildings of occupancy categories I, II, and, III in areas expecting SEVERE ground shaking.
    Seismic Design Category D are buildings of **all** occupancy groups in areas expecting severe and destructive ground shaking, but NOT located close to a major fault. (Sites with poor soil are a good example.)

Seismic Design Category E are buildings of occupancy groups I, II, and III in areas near major active faults. (Soil or rock is of no consequence.)
Seismic Design Category F are buildings of Occupancy Groups IV in areas near major active faults. (Soil or rock is of no consequence.)

The following are incorrect answers:
- Buildings in SDC A need to meet more restrictive earthquake requirements than Buildings in SDC C.
- SDC includes categories A through D.
- SDC includes categories A through G.

59. Answer: 1) In California, suspended ceilings must have diagonal bracing wires within the attic space, connecting the metal grid to the structure above. One set of four brace wires should be spaced 12 feet on center each way.

    The following are incorrect answers:
    - 6 feet on center each way
    - 12 feet on center
    - 6 feet on center

    See page 5 and page 9 of the FREE PDF file for ***Non-structural Earthquake Hazards in California Schools*** at the following link:
    http://www.documents.dgs.ca.gov/dsa/pubs/SB1122.pdf

60. Answer: 1) Glass adjacent to or within entrance or exit doors must be safety glass (i.e. tempered or laminated).

    The following are incorrect answers:
    - safety glass (i.e. tempered or wired)
    - safety glass (i.e. tempered or diagonal wired)
    - safety glass (i.e. tempered or annealed)

    Wired glass is no longer allowed by code.

    See page 20 of the FREE PDF file for ***Non-structural Earthquake Hazards in California Schools*** at the following link:
    http://www.documents.dgs.ca.gov/dsa/pubs/SB1122.pdf

61. Answer: 1) Per the CBC, curb ramps shall have a detectable warning that extends *the full width and depth* of the curb ramp, excluding the flared sides, inside the grooved border.

    The following are incorrect answers:
    - extends the full width and depth of the curb ramp, including the flared sides, inside the grooved border
    - extends the full width and two feet depth of the curb ramp, excluding the flared sides, inside the grooved border

- extends the full width and three feet depth of the curb ramp, excluding the flared sides, inside the grooved border

See CBC, Section 1127B.5 Curb ramps, Item 7. Detectable warnings, at the following link:
http://publicecodes.cyberregs.com/st/ca/st/b200v10/st_ca_st_b200v10_11b_sec029.htm?bu2=undefined

Note: The CBC detectable warning requirement is different from ADA requirements. ADA only requires that "detectable warning surfaces shall extend *610 mm (24 in) minimum in the direction of travel* and the full width of the curb ramp (exclusive of flares), the landing, or the blended transition."

See the link:
http://www.detectable-warning.com/guidelines.shtml

62. Answer: 4) Per the CBC, all of the following are acceptable dimensions for an accessible show stall:
    1) 30" x 60"
    2) 36" x 60"
    3) 42" x 48"

    See CBC Section 1127A.5.3, Showers, at the following link:
    http://publicecodes.cyberregs.com/st/ca/st/b200v10/st_ca_st_b200v10_11a_par239.htm

    Please note that the dimensions for accessible showers are not exactly the same between CBC and ADA. See the following link for ADA information:
    http://www.ada.gov/regs2010/2010ADAStandards/2010ADAstandards.htm

63. Answer: 2)
    Per the CBC, with a required occupant capacity of 100, the minimum corridor width shall not less than 44".

    The following are incorrect answers:
    - 36"
    - 48"
    - 60"

    Per CBC, Section 1018.2, Corridor width:
    "The minimum corridor width shall be as determined in Section 1005.1, but not less than 44 inches (1118 mm). Exceptions:
    - Twenty-four inches (610 mm)—For access to and utilization of electrical, mechanical or plumbing systems or equipment.
    - Thirty-six inches (914 mm)-With a required occupant capacity of less than 50.
    - Thirty-six inches (914 mm)-Within a dwelling unit.

- Seventy-two inches (1829 mm)-In Group E with a corridor having a required capacity of 100 or more.
- Seventy-two inches (1829 mm)-In corridors and areas serving gurney traffic in occupancies where patients receive outpatient medical care, which causes the patient to be not capable of self-preservation.
- Ninety-six inches (2438 mm)-In Group I-2 in areas where required for bed movement or corridors in Group I-2 and I-3 occupancies serving any area caring for one or more non-ambulatory persons."

See the following link:
http://publicecodes.cyberregs.com/st/ca/st/b200v10/st_ca_st_b200v10_10_sec018.htm

64. Answer: 3)
Per the CBC, the following is true:
- Ramps serving Group R occupancies may be 36 inches clear width when the occupant load is 50 or less.

The following are incorrect answers:
- Ramps shall have a minimum clear width of 44 inches, unless required to be wider by some other provision of this code.
- Ramps serving entrances to buildings, where the ramp is the only exit discharge path and serves an occupant load of 500 or more, shall have a minimum clear width of 60 inches.
- All of the above

Per CBC Section 1133B.5.2, Width:
"Ramps shall have a minimum clear width of 48 inches (1219 mm), unless required to be wider by some other provision of this code. Ramps serving entrances to buildings where the ramp is the only exit discharge path and serves an occupant load of 300 or more shall have a minimum clear width of 60 inches (1524 mm). Ramps serving Group R occupancies may be 36 inches (914 mm) clear width when the occupant load is 50 or less."

See the following link:
http://publicecodes.cyberregs.com/st/ca/st/b200v10/st_ca_st_b200v10_11b_sec035.htm

65. Answer: 4) None of the above

The following are incorrect answers:
- buildings of Occupancy Group U
- buildings that are unconditioned
- low-rise residential buildings that are heated with a wood heater or another non-mechanical heating system.

Per CEC Section 100 (a), Scope:
"Buildings covered. The provisions of Title 24, Part 6, apply to all buildings:

- That are of Occupancy Group A, B, E, F, H, M, R, S or U; and

- For which an application for a building permit or renewal of an existing permit is filed (or is required by law to be filed) on or after the effective date of the provisions, or which are constructed by a governmental agency; and

- That are:
  1) Unconditioned, indirectly or directly conditioned by mechanical heating or mechanical cooling or process spaces; or

  2) Low-rise residential buildings that are heated with a wood heater or another non-mechanical heating system.

Exception 1 to Section 100 (a): Qualified historic buildings as regulated by the California Historic Building Code (Title 24, Part 8).

Exception 2 to Section 100 (a): Building departments, at their discretion, may exempt temporary buildings, temporary outdoor lighting or temporary lighting in an unconditioned building, or structures erected in response to a natural disaster. Temporary buildings or structures shall be completely removed upon the expiration of the time limit stated in the permit."

See the following link:
http://publicecodes.cyberregs.com/st/ca/st/b1900v10/st_ca_st_b1900v10_1_sec001.htm

66. Answer: 2) Per the CEC, only habitable floors that have at least <u>50 percent</u> of their volume above grade as defined in the CBC shall be counted in determining how many habitable stories a building has.

    See the following link:
    http://publicecodes.cyberregs.com/st/ca/st/b1900v07/st_ca_st_b1900v07_subch1_sec001.htm

67. Answer: 3) In California, plans for a hospital building are submitted to the following governing agency for a plan check:
    - the Office of Statewide Health Planning and Development

    The following are incorrect answers:
    - the city
    - the health department
    - all of the above

See California Health and Safety Code, Section 127000-127050 at the following link: http://www.leginfo.ca.gov/cgi-bin/displaycode?section=hsc&group=126001-127000&file=127000-127050

AND
California Health and Safety Code, Section 129675-129680 at the following link: http://www.leginfo.ca.gov/cgi-bin/displaycode?section=hsc&group=129001-130000&file=129675-129680

68. Answer: 1) Per the California Health and Safety Code (CHSC), the following is required for a public pool:
    - an anti-entrapment grate

    The following are incorrect answers:
    - a backup safety system (This is not mandatory.)
    - an approved safety pool cover (This is one of the choices for a pool in single-family home.)
    - all of the above

    See California Health and Safety Code, Section 115920-115929 at the following link: http://www.leginfo.ca.gov/cgi-bin/displaycode?section=hsc&group=115001-116000&file=115920-115929
    http://leginfo.legislature.ca.gov/faces/billNavClient.xhtml?bill_id=201120120AB2114

69. Answer: 1) If a mechanics lien is filed against a property, the property owner is ultimately responsible for payment — even if she has already paid the prime contractor.

    The following are incorrect answers:
    - is no longer responsible for payment — as long as she has already paid the prime contractor
    - is no longer responsible for payment — as long as she can show proof that she has already paid the prime contractor
    - none of the above

    A lien can result when the prime contractor (referred to as a "direct contractor" in the mechanics lien revision statutes, effective July 1, 2012) has not paid subcontractors, laborers, or suppliers. Legally, the homeowner is ultimately responsible for payment — even if they already have paid the direct contractor. See the following link: http://www.cslb.ca.gov/Consumers/LegalIssuesForConsumers/MechanicsLien/WhatIsAMechanicsLien.asp

70. Answer: 1) A subcontractor or material supplier has <u>20</u> days after beginning work or delivering materials to serve the property owner a Preliminary Notice.

    See the following link:
    http://www.cslb.ca.gov/Consumers/LegalIssuesForConsumers/MechanicsLien/IfMechanicsLienFiledAgainstYou.asp

71. Answer: 2) The California Coastal Zone generally extends seaward to the state's outer limit of jurisdiction, including all offshore islands, and extends inland generally 1,000 yards from the mean high tide line of the sea.

    See Section 30103 of the California Coastal Act at the following link:
    http://www.coastal.ca.gov/coastact.pdf

72. Answer: 3) The California Clean Air Act was signed into law in 1988.

    See the following link:
    http://www.calepa.ca.gov/Legislation/1996/ab3048.htm

73. Answer: 1) The Field Act requires a base shear calculation, and that school buildings must be able to withstand lateral forces equal to at least 3% of the building's total mass.

    See the following link:
    http://en.wikipedia.org/wiki/Field_Act

74. Answer: 3) The California Senate SB 1953 sets a series of deadlines for compliance with the earlier Hospital Facilities Seismic Safety Act (HFSSA) of 1973. It requires that by the year 2008 all general acute-care hospitals must be able to withstand a major earthquake and all nonstructural mechanical, electrical, and plumbing systems in critical-care areas must be braced and anchored to assure survivability.

    By the year 2030 all general *acute-care inpatient* buildings, including approximately 470 general acute-care hospitals and 2,673 structures, must be in substantial compliance with the HFSSA.

    The following are incorrect answers:
    - all new general acute-care hospitals
    - all new general acute-care inpatient buildings
    - all general acute-care inpatient buildings

75. Answer: 4) The "Public entity," as used in the Public Contract Code, means the state, county, city, city and county, district, public authority, public agency, municipal corporation, or any other political subdivision or public corporation in the state.

    See the following link:
    http://www.leginfo.ca.gov/cgi-bin/displaycode?section=pcc&group=01001-02000&file=1100-1104

76. Answer: 1) The following statement regarding the values of California's wetlands is not accurate:
    - Biological Diversity: Wetlands provide important habitats for diverse communities of plants and animals, including about 30 percent of the federally listed threatened or endangered species. (It should be over 50 percent.)

See the following link:
http://ceres.ca.gov/ceres/calweb/wetlands/wetland_values.html

77. Answer: 4) Coastal California, also known as the California Coastline and the Golden Coast, refers to the coastal regions of the U.S. state of California. The term is based primarily on the following attributes EXCEPT geography.

The following are incorrect answers:

- culture
- economy
- politics

See the following link:
http://en.wikipedia.org/wiki/Coastal_California

78. Answer: 2) Per Sections 2081(b) and (c) of the California Endangered Species Act, if a State listed threatened and endangered species is found on a project site, the architect should assist the owner to obtain **an incidental take permit**.

The following are incorrect answers:

- Assist the owner to obtain a variance.
- Assist the owner to prepare an EIR.
- Assist the owner to obtain a CUP.

Sections 2081(b) and (c) of the California Endangered Species Act allow the California Department of Fish and Game to issue **an incidental take permit** for a State listed threatened and endangered species only if specific criteria are met. These criteria are reiterated in Title 14 CCR, Sections 783.4(a) and (b), and are as follows:

- The authorized take is incidental to an otherwise lawful activity;
- The impacts of the authorized take are minimized and fully mitigated;
- The measures required to minimize and fully mitigate the impacts of the authorized take:
  a) are roughly proportional in extent to the impact of the taking on the species,
  b) maintain the applicant's objectives to the greatest extent possible, and
  c) are capable of successful implementation;
- Adequate funding is provided to implement the required minimization and mitigation measures and to monitor compliance with and the effectiveness of the measures; and
- Issuance of the permit will not jeopardize the continued existence of a State-listed species

See the following link:
http://www.dfg.ca.gov/habcon/cesa/incidental/incid_perm_proced.html

79. Answer: 1) Clay is the most expansive.

    The following are incorrect answers:
    - silt
    - gravel
    - sand

80. Answer: 4) all of the above

    Soil liquefaction is "a phenomenon whereby a saturated or partially saturated soil substantially loses strength and stiffness in response to an applied stress, usually earthquake shaking or other sudden change in stress conditions, causing it to behave like a liquid."

    See the following link:
    http://en.wikipedia.org/wiki/Soil_liquefaction

    All of following can cause soil liquefaction:
    - standing underground water
    - an underground spring
    - earthquakes

81. Answer: 1) The following is an effective way to prevent soil erosion and flooding:
    - silt fencing

    The following are incorrect answers:
    - using a concrete patio around the building
    - maintaining a 2% slope away from the building
    - all of the above

    CAB is quietly starting to include general info on LEED and green buildings in the CSE without mentioning the name LEED. Just read the latest Test Plan of CSE and you will see.

    It is to your advantage to pick one or two good books on LEED and read them.

    Our other book, *LEED Green Associate Exam Guide* (ISBN: 9780984374137), can give you a very good overall knowledge on LEED, sustainability, and green buildings.

    All our books are available at:
    GreenExamEducation.com

82. Answer: 1) The *minimum* ventilation rate for an office per ASHRAE is 5 cfm/person.

    See the following link:
    http://hei-ohio.com/minimum%20ventilation%20rates.pdf

83. Answer: 4) Ventilation is the best way to control surface vapor pressure and related mold.

 The following cannot lower moisture levels and are incorrect answers:
 - humidifiers
 - HEPA filters
 - cooling the room

84. Answer: 4) During programming, an architect will typically consider inflation in an initial project cost estimate.

 The following items are too early to consider during the programming stage and are incorrect answers:
 - cost of utilities
 - labor cost
 - materials cost

85. Answer: 4) Both of the following will trigger a special inspection:
 - a field weld
 - using concrete with a strength of 3,000 psi or more *per the structural engineer's recommendation*

 Using concrete with a strength of 3,000 psi or more for better quality *per the architect's request* will NOT trigger a special inspection.

 Sometimes a city may require the use of concrete with a strength of 3,000 psi or more for footings in high-sulfate soils. This will NOT trigger a special inspection either since it is NOT required for the structural engineering.

86. Answer: 2) The following is the *minimum* qualification to be an inspector for Essential Service Buildings:
 - an inspector certified by ICBO (International Conference of Building officials).

 According to the Essential Services Building Act for the State of California, an **"essential services building"** means any building (or portion of building), including buildings designed and constructed for public agencies that are used, or designed to be used, as *a fire station, police station, emergency operations center, California Highway Patrol office, sheriff's office, or emergency communication dispatch center.*

 The following are incorrect answers:
 - an inspector certified by DSA (DSA stands for Division of the State Architect. This certificate is related to schools in California, and is above and beyond the *minimum* qualification. See the following link: http://www.ehow.com/info_10020164_dsa-certification.html.)
 - an inspector certified by OSHPD (OSHPD stands for Office of Statewide Health Planning and Development. This certificate is related to hospitals in California, and is above and beyond the *minimum* qualification.)

- an inspector certified by Essential Service Buildings Department of California (This certificate does not exist.)

87. Answer: 4) Per the energy code, a signed Installation Certificate should be
    - posted at the jobsite
      AND
    - provided to the building owner at occupancy

    See the following link:
    http://www.energy.ca.gov/2008publications/CEC-400-2008-016/rev1_appendices/appendix_A_files/Installation_Certificate_CF-6R-MECH-HERS/2008_CF-6R-MECH-25-HERS-RefrigerantCharge.pdf

    Note: The above link is pretty long. You can also go to the root directory (http://www.energy.ca.gov), and search with " CF-6R" to find the form.

88. Answer: 3) Per the energy code, a HERS rater should verify the refrigerant charge.

    HERS means Home Energy Rating System (HERS) Program. **HERS raters** are special inspectors certified by the California Energy Commission (CEC) to rate homes in California according to the Home Energy Rating System (HERS). These ratings include diagnostic testing and field verification to determine energy efficiency levels among homes tested for envelope leakage, building insulation, duct efficiency, and refrigerant charge verification for compliance with current building efficiency standards.

    See Form "CF-6R-MECH-25-HERS, Installation Certificate " at the previous link.

89. Answer: 3) Variances are used to
    - allow a property to be constructed in a manner basically consistent with the established zoning regulations with minor variations.

    A **variance** is a *limited* exception to the usual requirements of local zoning.

    The following are incorrect answers:
    - change design guidelines of a city
    - change the local zoning guidelines of a city
    - allow for a use that is completely different from the established zoning regulations

90. Answer: 4) The following is *not* true regarding a Conditional Use Permit (CUP):
    - A CUP allows special uses in a manner basically consistent with the established zoning regulations with minor variations.

    A **conditional use permit (CUP)** allows a local agency to consider special uses through a public hearing process. Such special uses are usually essential or *desirable* to a particular community, but *not allowed by zoning ordinances*.

Pay attention to the word "not." We are looking for an incorrect statement.

The following are correct statements but incorrect answers:
- A CUP usually requires a public hearing.
- A CUP is typically approved by a city or county.
- Any city or county zoning ordinance can list examples of common uses allowed with a conditional use permit.

91. Answer: 3) The cost of making an existing building accessible is 30% of the overall renovation construction cost. An architect should advise the owner about an unreasonable hardship waiver, and the potential lawsuits because of the non-compliance.

    Meeting the 20% construction cost hardship waiver does NOT protect the owner from a lawsuit under the **Unruh Civil Right Act**. The owner can still be sued for $4,000 or more per violation.

    See the link:
    http://sanbruno.ca.gov/comdev_images/buildingbrochures/ADA-Hardship.pdf

    The following are incorrect answers:
    - Advise the owner to make the building accessible.
    - Advise the owner to request an unreasonable hardship waiver.
    - Advise the owner about an unreasonable hardship waiver, and inform the owner that as long as the governing agencies approve the waiver, there is no potential for lawsuits for the non-compliance of laws.

92. Answer: 2) Per ADA, the clear floor or ground space for a wheel chair is 30"x48".

    See ADA, Section 305 at the following link:
    http://www.ada.gov/regs2010/2010ADAStandards/2010ADAStandards_prt.pdf

93. Answer: 1) The following is not related to the National Environmental Policy Act (NEPA)
    - Environmental Impact Report (EIR)

    NEPA uses the term Environmental Impact Statement (EIS) instead of Environmental Impact Report (EIR).

    EIR is required by CEQA.

    See the following links:
    http://en.wikipedia.org/wiki/National_Environmental_Policy_Act
    http://www.dfg.ca.gov/habcon/ceqa/intrnlproced/eir.html

    Pay attention to the word "not." We are looking for an incorrect statement.

The following are related to the National Environmental Policy Act (NEPA):

- Categorical Exclusion (CE)
- Finding of No Significant Impact (FONSI)
- Council on Environmental Quality (CEQ)

94. Answer: 2) The following is not a term used in FAA determinations:
    - denied

    Pay attention to the word "not." We are looking for an incorrect statement.

    The following are terms used in FAA determinations:
    - no objection
    - conditional approval
    - ALP ("An **ALP** is a graphic depiction of the existing and future airport facilities showing the clearance and dimensional requirements to meet applicable standards.")

    See the link:
    http://www.faa.gov/air_traffic/publications/atpubs/AIR/air1201.html

95. Answer: 1) ANSI is actively engaged in ISO 14000 (environmental).

    ANSI is the American National Standards Institute.

    See the link:
    http://ansi.org/about_ansi/overview/overview.aspx?menuid=1

    "**ISO 14000** is a *family of standards* related to environmental management that exists to help organizations (a) minimize how their operations (processes etc.) negatively affect the environment (i.e. cause adverse changes to air, water, or land); (b) comply with applicable laws, regulations, and other environmentally oriented requirements, and (c) continually improve in the above."

    See the link:
    http://en.wikipedia.org/wiki/ISO_14000

    The following are incorrect answers:
    - **ASTM is** ASTM International, formerly known as the American Society for Testing and Materials (ASTM).

      See the link:
      http://www.astm.org/

    - **Factory Mutual** is an "organization of a group *of insurers* composed of mutual property and casualty insurance companies, a subsidiary stock insurance company, and a subsidiary safety engineering company."

See the links:
http://www.answers.com/topic/factory-mutual
http://www.fmglobal.com/page.aspx?id=01000000

- **UL** "is a global independent *safety science* company offering expertise across seven key strategic businesses: Product Safety, Environment, Information and Insights..."

  See the link:
  http://www.ul.com/global/eng/pages/

96. Answer: 3) California has nine Regional Water Quality Control Boards (Regional Boards).

    See the link:
    http://www.swrcb.ca.gov/about_us/water_boards_structure/index.shtml

97. Answer: 1) For bathrooms and janitor rooms in retail buildings that process food or sell pre-packaged foods, environmental health departments will typically accept the following EXCEPT a floor with VCT flooring and a 6" continuous vinyl base.

    Pay attention to the word "EXCEPT." We are looking for an unacceptable design.

    Environmental health departments typically accept floor finishes that are washable and easy to clean. **VCT** is **vinyl composition tile** and not easy to clean because of the gaps between the vinyl tiles.

    The following are acceptable designs but incorrect answers:
    - a **ceramic tile floor** with a 6" high continuous ceramic tile cove base
    - a **sealed concrete floor** and a 6" high ceramic tile with a slim foot base
    - a **sheet vinyl floor** with a 6" continuous wrap-up base

98. Answer: 4) For a project that has impact on the nation's aquatic resources, the U.S. Army Corps of Engineers will review the application and grant a permit.

    See the link:
    http://www.usace.army.mil/Missions/CivilWorks/RegulatoryProgramandPermits.aspx

    The following are incorrect answers:
    - Coastal Commission
    - EPA (U.S. Environmental Protection Agency)
    - FEMA (Federal Emergency Management Agency)

99. Answer: 1) If there is a conflict between ADA and CBC, an architect should follow the most restrictive requirement.

100. Answer: 4) The following is a responsibility of a civil engineer in the schematic design stage:
- SWPPP (stormwater pollution prevention plan)

The following are incorrect answers:
- boundary survey and topography survey (This is done BEFORE the schematic design stage.)
- schematic parking lot layout (This is a responsibility of an architect.)
- schematic planting plan (This is a responsibility of a landscape architect.)

101. Answer: 1) After setting a date for your presentation to the Design Review Board (DRB) for your project, the Planning Department notifies residents adjacent to your project site of the date for your presentation to DRB so that the residents can attend the presentation and comment on your project.

The following are incorrect answers:
- notifies all city departments to solicit their comments on your project (Only a few people in the city are involved with the DRB review, not all city departments.)
- notifies all other governing agencies to solicit their comments on your project (These agencies will be contacted later.)
- forwards your project to the fire department for comments (This is not true; your project is typically not forwarded to fire department for comments at this early stage.)

102. Answer: 4) Typically a landscape architect will contribute documents for the presentation to the Design Review Board (DRB).

The following people will not contribute documents at this early stage, and are incorrect answers:
- electrical engineer
- mechanical engineer
- plumbing engineer

103. Answer: 2) The environmental health department typically requires a separate submittal because it is a county agency instead of a city agency.

The following are typically submitted together, and are incorrect answers:
- building department
- fire department (Some smaller cities do not have a fire department, and *sometimes* will require a separate submittal to the county's fire authority.)
- planning department

104. Answer: 4) If a client wants an architect to guarantee the highest standard of architectural service in the contract, the architect should inform the client that s/he can only provide an architectural service that is *ordinarily* provided by architects practicing under the *same or similar* circumstances.

See Section "Standard of Care" in B101–2007 (Former B141–1997), Standard Form of Agreement Between Owner and Architect (RAIC Document 6).

The following are incorrect answers:

- Incorporate the client's request in the contract.
- Negotiate with the client for a higher fee to guarantee the highest standard of architectural service.
- Inform the architect's insurance company about the client's request.

105. Answer: 4) all of the above

All the options listed are acceptable. Please note that an architect does NOT have to carry professional liability insurance to practice in California.

106. Answer: 3) **CalGreen** (California Green Building Code) requires construction waste reduction of at least 50%. Excavated soil and land clearing debris is exempt.

See CalGreen, Section 4.408 at the following link:
http://www.documents.dgs.ca.gov/bsc/CALGreen/2010_CA_Green_Bldg.pdf

107. Answer: 3) CalGreen requires a waste management plan for permit.

See CalGreen, Section 4.408 at the following link:
http://www.documents.dgs.ca.gov/bsc/CALGreen/2010_CA_Green_Bldg.pdf

The following are incorrect answers:

- ASHRAE calculations
- energy management plan
- water conservation plan

108. Answer: 2) CBC Chapter 17 requires a schedule of special inspections as a part of the permit submittal package.

See CBC Chapter 17 at the following link:
http://publicecodes.cyberregs.com/st/ca/st/b200v10/index.htm
The following are incorrect answers:

- a soils report
- structural calculations
- all of the above

109. Answer: 2) A 1,000-lb condensing unit on a vibration isolator does not need to be braced in a seismic zone.

Equipment on a vibration isolator does not need to be braced in a seismic zone.

The following are incorrect answers:
- a 2,000-lb air handler
- a 200-lb furnace
- a 30-lb pump mounted five feet above the floor

110. Answer: 1) Galvanized steel trapeze brackets are used to anchor conduits and pipes.

See the following link:
http://www.pexsupply.com/Holdrite-EZ-2-23-40-Galvanized-Steel-Trapeze-Bracket

The following are incorrect answers:
- recessed ceiling lights
- bookshelves
- water heaters

111. Answer: 2) once a week for at least two consecutive weeks or once a week for more than two consecutive weeks if the longer period of advertising is deemed necessary by the department

See Public Contract Code, Section 10140-10141 at the following link:
http://www.leginfo.ca.gov/cgi-bin/displaycode?section=pcc&group=10001-11000&file=10140-10141

112. Answer: 4) At a job site visit, an architect notices that the roofing contractor is not following safety procedures. The architect should inform the general contractor and the owner in writing.

The following are incorrect answers:
- Tell the roofing contractor to follow the safety procedures. (An architect should NEVER direct a sub-contractor; all communications to sub-contractors should pass through general contractor.)
- Tell the roofing contractor to follow the safety procedures and also inform the general contractor.
- Inform the general contractor in writing. (This will not protect the architect; the architect should also inform the owner since the architect has a contract with the owner.)

113. Answer: 1) If the construction is delayed by a fire at the job site, the architect should issue a change order to extend the contract time as determined by the architect to be reasonable.

See A201–2007, General Conditions of the Contract for Construction, Section 8.3, Delays and Extension of Time.

114. Answer: 1) During construction of a renovation project, the contractor discovers asbestos in the ceiling and attic. The contractor should stop the construction, and inform the owner and architect in writing. The owner should hire specialists to test and render the asbestos harmless at the owner's cost.

See A201–2007, General Conditions of the Contract for Construction, Section 10.3, Delays and Extension of Time.

115. Answer: 1) The most likely sequence to resolve disputes between an owner and contractor is initial decision, mediation, arbitration, and litigation.

See A201–2007, General Conditions of the Contract for Construction, Section 15, Claims and Disputes.

116. Answer: 4) During construction, minor changes do not affect contract time or contract sum of the project if issued as an architect's supplemental instruction.

See A201–2007, General Conditions of the Contract for Construction, Sections 7.3 and 7.4.

The following are incorrect answers:

- addendum (These are issued before the bid due date, not during construction.)
- change order (These are for changes affecting contract time or contract sum during construction.)
- construction change directive (These are for minor changes that may affect contract time or contract sum, and happen before the owner and contractor agree on the cost.)

117. Answer: 1) The schedule of values is determined by the contractor, subject to the architect's review.

See A201–2007, General Conditions of the Contract for Construction, Sections 9.2, Schedule of Values.

118. Answer: 2) An architect notices that the contractor's application for payment includes the materials stored at the site and materials stored off the site. The architect should approve payment for the materials stored at the site and complete more research on payment for the materials stored off the site, since these materials might qualify for payment also.

The architect should definitely approve payment for the materials stored at the site, and he should approve materials stored off the site *only* if it is approved *in advance* by the owner.

See A201–2007, General Conditions of the Contract for Construction, Sections 9.3, Application for Payment.

119. Answer: 1) Warranties required by the contract documents shall *typically* commerce on the date of substantial completion.

Pay attention to the word "typically." Of course, there are always exceptions: warranties for items not completed and still on the punch list will not commence until they are finished.

See A201–2007, General Conditions of the Contract for Construction, Sections 9.8, Substantial Completion.

120. Answer: 3) A sub-contractor is required to file a preliminary notice to be able to record a lien.

See the following link:
http://www.cab.ca.gov/pdf/misc/mechanics_lien_law.pdf

C. How We Come Up with the CSE Mock Exam Questions

We come up with all the CSE Mock Exam questions based on the CSE Test Plan, and we make sure we include at least two questions for each of the 55 CSE knowledge statements.

Based on CSE Test Plan, we developed the CSE Mock Exam based on the *four* weighted primary knowledge categories:

Categories and Sub-Categories	**Weight/Number of Questions Expected**	**Actual**
1) Context and Pre-Design (Knowledge Statements 1 to 8)	16% or 19	17% or 20
2) Regulatory	42% or 51	45% or 54
a) California State Laws, Code, Regulations, and Standards (Knowledge Statements 9 to 23)	24% or 30	28% or 33
b) Other Laws, Code, Regulations, Standards, Agencies and Entitlements (Knowledge Statements 24 to 33)	18% or 21	18% or 21
3) Management and Design (Knowledge Statements 34 to 46)	27% or 32	22% or 26
4) Construction (Knowledge Statements 47 to 55)	15% or 18	17% or 20
Total	100% or 120	120

See a detailed breakdown in the following tables:

Note: *If the text on following tables is too small for you to read, then you can go to our forum, sign up for a free account, and download the FREE 11x17 full size jpeg format files for these tables at:*
GeeForum.com

CSE Mock Exam Statistics by Gang Chen, GreenExamEducation.com								
					2) Regulatory 42% or 42			
Knowledge Statement	Description	# of Q	Actual Q #	1) Context and Pre-design (Knowledge Statements 1 to 8) 16% or 19	2a) California State Laws, Code, Regulations, and Standards (Knowledge Statements 9 to 23) 24% or 30	2b) Other Laws, Code, Regulations, Standards, Agencies and Entitlements (Knowledge Statements 24 to 33) 18% or 21	3) Management and Design (Knowledge Statements 34 to 46) 27% or 32	4) Construction (Knowledge Statements 47 to 55) 15% or 18
1	Natural environment (wetlands, coastal regions, habits of endangered species)	4	76, 77, 78, 40	4				
2	Natural & human-caused hazardous condition (seismic activities, winds, flood zone, fire, hazardous matrials, etc,) & potential mitigation	4	79, 80, 114, 40	4				
3	Health issues (off gasing, mold, adequate ventilation)	2	82, 83	2				
4	User characteristics (varying ages, cultures, abilities, activities requirements)	2	5, 9	2				
5	Types of stakeholders (environmental groups, citizen's advisory committees, neighborhood and community organizations)	2	7, 8	2				
6	Project performances objectives (comfort control, safety and security, sustainability)	2	81, 4	2				
7	Evaluate user activities for adjacencies & spatial requirements	2	6, 10	2				
8	Organize and evaluate program documents	2	84, 11	2				
9	California Environmental Quality Act (CEQA)	2	51, 52		2			
10	Essential Services Buildings Seismic Safety Act	2	53, 54		2			
11	CBSC components	2	55, 56		2			
12	CBC structural provision	2	57, 58		2			
13	CBC anchoring and bracing non-structural elements	2	59, 60		2			
14	CBC accessibility provision	2	61, 62		2			
15	CBC fire & life safety	2	63, 64		2			
16	California Energy Code (CEC)	2	65, 66		2			
17	California Health and Safety Code (CHSC)	2	67,68		2			
18	Mechanic's Lien	2	69, 70		2			
19	CA special regulations (water regulations, CA Coastal Act, CA Clean Air Act, CA Public Contract Code, Field Act, Hospital Facilities Seismic Safety Act)	5	71, 72, 73, 74, 75		5			
20	Architect Practice Act (APA)	2	47, 48		2			
21	APA	2	44, 49		2			
22	APA	2	46, 50		2			
23	APA	2	43, 45		2			

Table 4.1a Breakdown of the Mock Exam Questions

CSE Mock Exam Statistics by Gang Chen, GreenExamEducation.com								
					2) Regulatory 42% or 42			
Knowledge Statement	Description	# of Q	Actual Q #	1) Context and Pre-design (Knowledge Statements 1 to 8) 16% or 19	2a) California State Laws, Code, Regulations, and Standards (Knowledge Statements 9 to 23) 24% or 30	2b) Other Laws, Code, Regulations, Standards, Agencies and Entitlements (Knowledge Statements 24 to 33) 18% or 21	3) Management and Design (Knowledge Statements 34 to 46) 27% or 32	4) Construction (Knowledge Statements 47 to 55) 15% or 18
24	Local codes & regulations (General plan, planning, zoning & building ordinances, guidelines, CC&R)	2	89, 90			2		
25	ADA	2	91, 92			2		
26	Other federal laws (FAA, EPA, US Army Corps of Engineers)	3	93,94, 98			3		
27	National standards (UL, ANSI, ASTM, Factory Mutual)	2	95, 13			2		
28	Local community development agencies (Planning, Public Works, Building, Fire)	2	86, 107			2		
29	Local or regional agencies (DRB, AQMD, County Flood Control District, airport authorities, Environmental Health)	2	97, 106			2		
30	State agencies (Coastal Commission, Water Resource Control Board, Dept of Fish and Game, Air Resources Board, CA DOT)	2	96, 14			2		
31	Procedure for obtaining agency approvals	2	101, 102			2		
32	Interrelationships among agencies	2	103, 15			2		
33	Process of resolving conflicts between agencies & codes	2	99, 17			2		
34	Consultants (CLSEMP, Acoustic, traffic)	2	100, 18				2	
35	Architect's role in coordinating the project team	2	19, 20				2	
36	Architect's role in managing the project team to get agency approvals	2	12, 16				2	
37	Quality assurance (In-house & external)	2	22, 23				2	
38	Standards of care & liabilities	2	104, 105				2	
39	Methods for developing design with stakeholders	2	21, 24				2	
40	Contents of drawings & documents for agency approvals	2	1, 2				2	
41	Non-structural elements (FFE, nonbearing partitions, suspended ceilings)	2	109, 110				2	
42	Special sturctural loading condition (heavy equipment, snow, library shelving)	2	108, 25				2	
43	Environmental control system (energy management, occupant comfort & control)	2	87, 88				2	
44	Material characteristics, performance, & testing standards.	2	3, 26				2	

Table 4.1b Breakdown of the Mock Exam Questions

CSE Mock								
					2) Regulatory 42% or 42			
Knowledge Statement	Description	# of Q	Actual Q #	1) Context and Pre-design (Knowledge Statements 1 to 8) 16% or 19	2a) California State Laws, Code, Regulations, and Standards (Knowledge Statements 9 to 23) 24% or 30	2b) Other Laws, Code, Regulations, Standards, Agencies and Entitlements (Knowledge Statements 24 to 33) 18% or 21	3) Management and Design (Knowledge Statements 34 to 46) 27% or 32	4) Construction (Knowledge Statements 47 to 55) 15% or 18
45	Ability to prepare CD per project type.	2	27, 28				2	
46	Bidding & negotiation	2	111, 29				2	
47	Interrelationships among owner, architect & GC	2	30, 31					2
48	Limits of architect's role during CA (directing sub, means & method)	2	112, 32					2
49	Conflicts resolutions during CA (mediation, arbitration, litigation)	2	115, 33					2
50	Changes during CA (CCD, CO, supplemental instruction)	2	113, 116					2
51	Monitoring costs & schedule (review & certify payments, review lien release)	2	117, 118					2
52	Project close-out (Certificate of Substantial Completion, Notice of Completion, Final Lien Releases)	4	119, 34, 41, 42					4
53	CA construction laws (lien requirements, min. warranty period)	2	120, 35					2
54	Special inspection (field welding, high strength conc)	2	85, 36					2
55	Ability to do CA per project type	2	37, 38					2
Subtotal				20	33	21	26	20
Total		120						

Table 4.1c Breakdown of the Mock Exam Questions

Appendixes

A. List of Figures

B. List of Tables

C. Official reference materials suggested by NCARB & CAB

1. General NCARB reference materials for ARE:

Per NCARB, all candidates should become familiar with the latest version of the following codes:

International Code Council, Inc. (ICC, 2006)
International Building Code
International Mechanical Code
International Plumbing Code

National Fire Protection Association (NFPA)
Life Safety Code (NFPA 101)
National Electrical Code (NFPA 70)

National Research Council of Canada
National Building Code of Canada
National Plumbing Code of Canada
National Fire Code of Canada

American Institute of Architects
AIA Documents - 2007

Candidates should be familiar with the Standard on Accessible and Usable Buildings and Facilities (ICC/ANSI A117.1-98)

2. Official CAB reference materials for CSE:

You need to check this website periodically for future updates:
http://www.cab.ca.gov/candidates/cse_materials.shtml

The following were updated on August 12, 2011

Reference Material	Source
B102, B201, and A201 of the 2007 American Institute of Architects (AIA) Contract Documents	American Institute of Architects (AIA)
California Architects Practice Act (Latest Edition)	California Architects Board
California Contractors License Law and Reference Book (Lien Laws)	Local County Law Library
California Public Contract Code (State Contract Act 10100 - 10285.5)	Local County Law Library or Leginfo

General Plan (Government Code, Section 65300 Et. Seq.)	Local County Law Library or Leginfo
Essential Services Buildings Seismic Safety Act, Excerpt (California Health & Safety Code, Sections 16000 - 16023)	Local County Law Library or Leginfo
California Design Professionals' Liens (California Civil Code, Sections 8300 - 8319)	Local County Law Library or Leginfo
California Coastal Act (Public Resources Code, Section 30000 Et. Seq.)	California Coastal Commission or Leginfo
AIA Architect's Handbook of Professional Practice	American Institute of Architects or American Institute of Architects California Council
2010 California Building Standards Administrative Code (Part 1, Title 24)*	California Building Standards Commission
2010 California Building Code, Volumes 1 & 2 (Part 2, Title 24)*	California Building Standards Commission
2010 California Residential Building Code (Part 2.5, Title 24)*	California Building Standards Commission
2010 California Electrical Code (Part 3, Title 24)*	California Building Standards Commission
2010 California Mechanical Code (Part 4, Title 24)*	California Building Standards Commission
2010 California Plumbing Code (Part 5, Title 24)*	California Building Standards Commission
2010 California Energy Code (Part 6, Title 24)*	California Building Standards Commission
2010 California Historical Building Code (Part 8, Title 24)*	California Building Standards Commission
2010 California Fire Code (Part 9, Title 24)*	California Building Standards Commission
2010 California Existing Building Standards Code (Part 10, Title 24)*	California Building Standards Commission
2010 California Green Building Standards Code (Part 11, Title 24)*	California Building Standards Commission
2010 California Referenced Standards Code (Part 12, Title 24)*	California Building Standards Commission
California Environmental Quality Act (Public Resources Code, Section 21000 Et. Seq.)	American Council of Engineering Companies or California Environmental Resources Evaluation System
California Coastal Zone Conservation Act (Public Resources Code, Division 20)	http://www.coastal.ca.gov/coastact.pdf

State of California Indoor Air Quality Program	State of California Indoor Air Quality Program
Minimum Design Loads for Buildings and Other Structures, 2005 ASCE 7-05	Various Sources
Local Government Regulations (i.e., design review standards, development project regulations, and/or ordinances)	Various Local Sources

* The Title 24 Parts listed above can also be found at your local library, county law library, or architectural college library. A listing of Title 24 depositories is available from the California Building Standards Commission.
Supplies may be limited and the time needed to process your request lengthy, please order items from sources as soon as possible.

CSE candidates should be familiar with agency regulations related to the purview of the Federal Aviation Administration, the US Army Corp of Engineers, and the Americans with Disabilities Act (as they pertain to architectural practice in California).

Candidates should also be knowledgeable about and understand the Mechanic's Lien Laws, and the use of the Notice of Completion.

D. Other reference materials

Chen, Gang. *Building Construction: Project Management, Construction Administration, Drawings, Specs, Detailing Tips, Schedules, Checklists, and Secrets Others Don't Tell You (Architectural Practice Simplified, 2nd edition).* ArchiteG, Inc., A good introduction to the architectural practice and construction documents and service, including discussions of MasterSpec format and specification sections.

Chen, Gang. *LEED GA Exam Guide: A Must-Have for the LEED Green Associate Exam: Comprehensive Study Materials, Sample Questions, Mock Exam, Green Building LEED Certification, and Sustainability.* ArchiteG, Inc., latest edition. A good introduction to green buildings and the LEED building rating system.

Ching, Francis. *Architecture: Form, Space, & Order.* Wiley, latest edition. It is one of the best architectural books that you can have. I still flip through it every now and then. It is a great book for inspiration.

Ching, Francis. Steven R. Winkel, FAIA, PE. *Building Codes Illustrated: A Guide to Understanding the International Building Code.* Wiley, latest edition. A valuable interpretive guide with many useful line drawings. A great timesaver.

Frampton, Kenneth. *Modern Architecture: A Critical History.* Thames and Hudson, London, latest edition. A valuable resource for architectural history.

Jarzombek, Mark M. (Author), Vikramaditya Prakash (Author), Francis D. K. Ching (Editor). *A Global History of Architecture.* Wiley, latest edition. A valuable and comprehensive resource for architectural history with 1000 b & w photos, 50 color photos, and 1500 b & w illustrations. It doesn't limit the topic on a Western perspective, but rather through a global vision.

Trachtenberg, Marvin and Isabelle Hyman. *Architecture: From Pre-history to Post-Modernism.* Prentice Hall, Englewood Cliffs, NJ latest edition. A valuable and comprehensive resource for architectural history.

E. Definition of Architects and Some Important Information about Architects and the Profession of Architecture

Architects, Except Landscape and Naval

- Nature of the Work
- Training, Other Qualifications, and Advancement
- Employment
- Job Outlook
- Projections Data
- Earnings
- OES Data
- Related Occupations
- Sources of Additional Information

Significant Points

- About 1 in 5 architects are self-employed—more than 2 times the proportion for all occupations.
- Licensing requirements include a professional degree in architecture, at least 3 years of practical work training, and passing all divisions of the Architect Registration Examination.
- Architecture graduates may face competition, especially for jobs in the most prestigious firms.

Nature of the Work

People need places in which to live, work, play, learn, worship, meet, govern, shop, and eat. These places may be private or public; indoors or out; rooms, buildings, or complexes, and architects design them. Architects are licensed professionals trained in the art and science of building design who develop the concepts for structures and turn those concepts into images and plans.

Architects create the overall aesthetic and look of buildings and other structures, but the design of a building involves far more than its appearance. Buildings also must be functional, safe, and economical and must suit the needs of the people who use them. Architects consider all these factors when they design buildings and other structures.

Architects may be involved in all phases of a construction project, from the initial discussion with the client through the entire construction process. Their duties require specific skills—designing, engineering, managing, supervising, and communicating with clients and builders. Architects spend a great deal of time explaining their ideas to clients, construction contractors, and others. Successful architects must be able to communicate their unique vision persuasively.

The architect and client discuss the objectives, requirements, and budget of a project. In some cases, architects provide various pre-design services: conducting feasibility and environmental impact studies, selecting a site, preparing cost analysis and land-use studies, or specifying the requirements the design must meet. For example, they may determine space requirements by researching the numbers and types of potential users of a building. The architect then prepares drawings and a report presenting ideas for the client to review.

After discussing and agreeing on the initial proposal, architects develop final construction plans that show the building's appearance and details for its construction. Accompanying these plans are drawings of the structural system; air-conditioning, heating, and ventilating systems; electrical systems; communications systems; plumbing; and, possibly, site and landscape plans. The plans also specify the building materials and, in some cases, the interior furnishings. In developing designs, architects follow building codes, zoning laws, fire regulations, and other ordinances, such as those requiring easy access by people who are disabled. Computer-aided design and drafting (CADD) and Building Information Modeling (BIM) technology has replaced traditional paper and pencil as the most common method for creating design and construction drawings. Continual revision of plans on the basis of client needs and budget constraints is often necessary.

Architects may also assist clients in obtaining construction bids, selecting contractors, and negotiating construction contracts. As construction proceeds, they may visit building sites to make sure that contractors follow the design, adhere to the schedule, use the specified materials, and meet work quality standards. The job is not complete until all construction is finished, required tests are conducted, and construction costs are paid. Sometimes, architects also provide post-construction services, such as facilities management. They advise on energy efficiency measures, evaluate how well the building design adapts to the needs of occupants, and make necessary improvements.

Often working with engineers, urban planners, interior designers, landscape architects, and other professionals, architects in fact spend a great deal of their time coordinating information from, and the work of, other professionals engaged in the same project.

They design a wide variety of buildings, such as office and apartment buildings, schools, churches, factories, hospitals, houses, and airport terminals. They also design complexes such as urban centers, college campuses, industrial parks, and entire communities.

Architects sometimes specialize in one phase of work. Some specialize in the design of one type of building—for example, hospitals, schools, or housing. Others focus on planning and pre-design services or construction management and do minimal design work.

Work environment. Usually working in a comfortable environment, architects spend most of their time in offices consulting with clients, developing reports and drawings, and working with other architects and engineers. However, they often visit construction sites to review the progress of projects. Although most architects work approximately 40 hours per week, they often have to work nights and weekends to meet deadlines.

Training, Other Qualifications, and Advancement

There are three main steps in becoming an architect. First is the attainment of a professional degree in architecture. Second is work experience through an internship, and third is licensure through the passing of the Architect Registration Exam.

Education and training. In most States, the professional degree in architecture must be from one of the 114 schools of architecture that have degree programs accredited by the National Architectural Accrediting Board. However, State architectural registration boards set their own standards, so graduation from a non-accredited program may meet the educational requirement for licensing in a few States.

Three types of professional degrees in architecture are available: a 5-year bachelor's degree, which is most common and is intended for students with no previous architectural training; a 2-year master's degree for students with an undergraduate degree in architecture or a related area; and a 3- or 4-year master's degree for students with a degree in another discipline.

The choice of degree depends on preference and educational background. Prospective architecture students should consider the options before committing to a program. For example, although the 5-year bachelor of architecture offers the fastest route to the professional degree, courses are specialized, and if the student does not complete the program, transferring to a program in another discipline may be difficult. A typical program includes courses in architectural history and theory, building design with an emphasis on CADD, structures, technology, construction methods, professional practice, math, physical sciences, and liberal arts. Central to most architectural programs is the design studio, where students apply the skills and concepts learned in the classroom, creating drawings and three-dimensional models of their designs.

Many schools of architecture also offer post-professional degrees for those who already have a bachelor's or master's degree in architecture or other areas. Although graduate education beyond the professional degree is not required for practicing architects, it may be required for research, teaching, and certain specialties.

All State architectural registration boards require architecture graduates to complete a training period—usually at least 3 years—before they may sit for the licensing exam. Every State, with the exception of Arizona, has adopted the training standards established by the Intern Development Program, a branch of the American Institute of Architects and the National Council of Architectural Registration Boards (NCARB). These standards stipulate broad training under the supervision of a licensed architect. Most new graduates complete their training period by working as interns at architectural firms. Some States allow a portion of the training to occur in the offices of related professionals, such as engineers or general contractors. Architecture students who complete internships while still in school can count some of that time toward the 3-year training period.

Interns in architectural firms may assist in the design of one part of a project, help prepare architectural documents or drawings, build models, or prepare construction drawings on CADD. Interns also may research building codes and materials or write specifications for building materials, installation criteria, the quality of finishes, and other, related details.

Licensure. All States and the District of Columbia require individuals to be licensed (registered) before they may call themselves architects and contract to provide architectural services. During the time between graduation and becoming licensed, architecture school graduates generally work in the field under the supervision of a licensed architect who takes legal responsibility for all work. Licensing requirements include a professional degree in architecture, a period of practical training or internship, and a passing score on all divisions of the Architect Registration Examination. The examination is broken into nine divisions consisting of either multiple choice or graphical questions. The eligibility period for completion of all divisions of the exam varies by State.

Most States also require some form of continuing education to maintain a license, and many others are expected to adopt mandatory continuing education. Requirements vary by State but usually involve the completion of a certain number of credits annually or biennially through workshops, formal university classes, conferences, self-study courses, or other sources.

Other qualifications. Architects must be able to communicate their ideas visually to their clients. Artistic and drawing ability is helpful, but not essential, to such communication. More important are a visual orientation and the ability to understand spatial relationships. Other important qualities for anyone interested in becoming an architect are creativity and the ability to work independently and as part of a team. Computer skills are also required for writing specifications, for 2- and 3- dimensional drafting using CADD programs, and for financial management.

Certification and advancement. A growing number of architects voluntarily seek certification by the National Council of Architectural Registration Boards. Certification is awarded after independent verification of the candidate's educational transcripts, employment record, and professional references. Certification can make it easier to become licensed across States. In fact, it is the primary requirement for reciprocity of licensing among State Boards that are NCARB members. In 2007, approximately one-third of all licensed architects had this certification.

After becoming licensed and gaining experience, architects take on increasingly responsible duties, eventually managing entire projects. In large firms, architects may advance to supervisory or managerial positions. Some architects become partners in established firms, while others set up their own practices. Some graduates with degrees in architecture also enter related fields, such as graphic, interior, or industrial design; urban planning; real estate development; civil engineering; and construction management.

Employment

Architects held about 132,000 jobs in 2006. Approximately 7 out of 10 jobs were in the architectural, engineering, and related services industry—mostly in architectural firms with fewer than five workers. A small number worked for residential and nonresidential building construction firms and for government agencies responsible for housing, community planning, or construction of government buildings, such as the U.S. Departments of Defense and Interior, and the General Services Administration. About 1 in 5 architects are self-employed.

Job Outlook

Employment of architects is expected to grow faster than the average for all occupations through 2016. Keen competition is expected for positions at the most prestigious firms, and opportunities will be best for those architects who are able to distinguish themselves with their creativity.

Employment change. Employment of architects is expected to grow by 18 percent between 2006 and 2016, which is faster than the average for all occupations. Employment of architects is strongly tied to the activity of the construction industry. Strong growth is expected to come from nonresidential construction as demand for commercial space increases. Residential construction, buoyed by low interest rates, is also expected to grow as more people become homeowners. If interest rates rise significantly, home building may fall off, but residential construction makes up only a small part of architects' work.

Current demographic trends also support an increase in demand for architects. As the population of Sunbelt States continues to grow, the people living there will need new places to live and work. As the population continues to live longer and baby-boomers begin to retire, there will be a need for more healthcare facilities, nursing homes, and retirement communities. In education, buildings at all levels are getting older and class sizes are getting larger. This will require many school districts and universities to build new facilities and renovate existing ones.

In recent years, some architecture firms have outsourced the drafting of construction documents and basic design for large-scale commercial and residential projects to architecture firms overseas. This trend is expected to continue and may have a negative impact on employment growth for lower level architects and interns who would normally gain experience by producing these drawings.

Job prospects. Besides employment growth, additional job openings will arise from the need to replace the many architects who are nearing retirement, and others who transfer to other occupations or stop working for other reasons. Internship opportunities for new architectural students are expected to be good over the next decade, but more students are graduating with architectural degrees and some competition for entry-level jobs can be anticipated. Competition will be especially keen for jobs at the most prestigious architectural firms as prospective architects try to build their reputation. Prospective architects who have had internships while in school will have an advantage in obtaining intern positions after graduation. Opportunities will be best for those architects that are able to distinguish themselves from others with their creativity.

Prospects will also be favorable for architects with knowledge of "green" design. Green design, also known as sustainable design, emphasizes energy efficiency, renewable resources such as energy and water, waste reduction, and environmentally friendly design, specifications, and materials. Rising energy costs and increased concern about the environment has led to many new buildings being built green.

Some types of construction are sensitive to cyclical changes in the economy. Architects seeking design projects for office and retail construction will face especially strong competition for jobs or clients during recessions, and layoffs may ensue in less successful firms. Those involved in the design of institutional buildings, such as schools, hospitals, nursing homes, and correctional facilities, will be less affected by fluctuations in the economy. Residential construction makes up a small portion of work for architects, so major changes in the housing market would not be as significant as fluctuations in the nonresidential market.

Despite good overall job opportunities, some architects may not fare as well as others. The profession is geographically sensitive, and some parts of the Nation may have fewer new building projects. Also, many firms specialize in specific buildings, such as hospitals or office towers, and demand for these buildings may vary by region. Architects may find it increasingly necessary to gain reciprocity in order to compete for the best jobs and projects in other States.

Projections Data

Projections data from the National Employment Matrix

Occupational title	SOC Code	Employment, 2006	Projected employment, 2016	Change, 2006-16		Detailed statistics	
				Number	Percent		
Architects, except landscape and naval	17-1011	132,000	155,000	23,000	18	PDF	zipped XLS

NOTE: Data in this table are rounded. See the discussion of the employment projections table in the *Handbook* introductory chapter on *Occupational Information Included in the Handbook*.

Earnings

Median annual earnings of wage-and-salary architects were $64,150 in May 2006. The middle 50 percent earned between $49,780 and $83,450. The lowest 10 percent earned less than $39,420, and the highest 10 percent earned more than $104,970. Those just starting their internships can expect to earn considerably less.

Earnings of partners in established architectural firms may fluctuate because of changing business conditions. Some architects may have difficulty establishing their own practices and may go through a period when their expenses are greater than their income, requiring substantial financial resources.

Many firms pay tuition and fees toward continuing education requirements for their employees.

For the latest wage information:

The above wage data are from the Occupational Employment Statistics (OES) survey program, unless otherwise noted. For the latest National, State, and local earnings data, visit the following pages:

Architects, except landscape and naval

Related Occupations

Architects design buildings and related structures. Construction managers, like architects, also plan and coordinate activities concerned with the construction and maintenance of buildings and facilities. Others who engage in similar work are landscape architects, civil engineers, urban and regional planners, and designers, including interior designers, commercial and industrial designers, and graphic designers.

Sources of Additional Information

Disclaimer:

Links to non-BLS Internet sites are provided for your convenience and do not constitute an endorsement.

Information about education and careers in architecture can be obtained from:

- The American Institute of Architects, 1735 New York Ave. NW., Washington, DC 20006. Internet: http://www.aia.org
- Intern Development Program, National Council of Architectural Registration Boards, Suite 1100K, 1801 K St. NW., Washington, D.C. 20006. Internet: http://www.ncarb.org OOH ONET Codes 17-1011.00"

Quoted from: Bureau of Labor Statistics, U.S. Department of Labor, Occupational Outlook Handbook, 2008-09 Edition, Architects, Except Landscape and Naval, on the Internet at **http://www.bls.gov/oco/ocos038.htm** (visited November 30, 2008).
Last Modified Date: December 18, 2007

Note: *Please check the website above for the latest information.*

F. AIA Compensation Survey

Every 3 years, AIA publishes a Compensation Survey for various positions at architectural firms across the country. It is a good idea to find out the salary before you make the final decision to become an architect. If you are already an architect, it is also a good idea to determine if you are underpaid or overpaid.

See following link for some sample pages for the latest AIA Compensation Survey:

https://www.aia.org/resources/8066-aia-compensation-report

G. So … You would Like to Study Architecture

To study architecture, you need to learn how to draft, how to understand and organize spaces and the interactions between interior and exterior spaces, how to do design, and how to communicate effectively. You also need to understand the history of architecture.

As an architect, a leader for a team of various design professionals, you not only need to know architecture, but also need to understand enough of your consultants' work to be able to coordinate them. Your consultants include soils and civil engineers, landscape architects, structural, electrical, mechanical, and plumbing engineers, interior designers, sign consultants, etc.

There are two major career paths for you in architecture: practice as an architect or teach in colleges or universities. The earlier you determine which path you are going to take, the more likely you will be successful at an early age. Some famous and well-respected architects, like my USC alumnus Frank Gehry, have combined the two paths successfully. They teach at the universities and have their own architectural practice. Even as a college or university professor, people respect you more if you have actual working experience and have some built projects. If you only teach in colleges or universities but have no actual working experience and have no built projects, people will consider you as a "paper" architect, and they are not likely to take you seriously, because they will think you probably do not know how to put a real building together.

In the U.S., if you want to practice architecture, you need to obtain an architect's license. It requires a combination of passing scores on the Architectural Registration Exam (ARE) and 8 years of education and/or qualified working experience, including at least 1 year of working experience in the U.S. Your working experience needs to be under the supervision of a licensed architect to be counted as qualified working experience for your architect's license.

If you work for a landscape architect or civil engineer or structural engineer, some states' architectural licensing boards will count your experience at a discounted rate for the qualification of your architect's license. For example, 2 years of experience working for a civil engineer may be counted as 1 year of qualified experience for your architect's license. You need to contact your state's architectural licensing board for specific licensing requirements for your state.

If you want to teach in colleges or universities, you probably want to obtain a master's degree or a Ph.D. It is not very common for people in the architectural field to have a Ph.D. One reason is that there are few Ph.D. programs for architecture. Another reason is that architecture is considered a profession and requires a license. Many people think an architect's license is more important than a Ph.D. degree. In many states, you need to have an architect's license to even use the title "architect," or the terms "architectural" or "architecture" to advertise your service. You cannot call yourself an architect if you do not have an architect's license, even if you have a Ph.D. in architecture. Violation of these rules brings punishment.

To become a tenured professor, you need to have a certain number of publications and pass the evaluation for the tenure position. Publications are very important for tenure track positions. Some people say for the tenured track positions in universities and colleges, it is "publish or perish."

The American Institute of Architects (AIA) is the national organization for the architectural profession. Membership is voluntary. There are different levels of AIA membership. Only licensed architects can be (full) AIA members. If you are an architectural student or an intern but not a licensed architect yet, you can join as an associate AIA member. Contact AIA for detailed information.

The National Council of Architectural Registration Boards (NCARB) is a nonprofit federation of architectural licensing boards. It has some very useful programs, such as IDP, to assist you in obtaining your architect's license. Contact NCARB for detailed information.

Back Page Promotion

You may be interested in some other books written by Gang Chen:

A. **ARE Mock Exam series. See the following link:**
   http://www.GreenExamEducation.com

B. **LEED Exam Guides series. See the following link:**
   http://www.GreenExamEducation.com

C. ***Building Construction:*** *Project Management, Construction Administration, Drawings, Specs, Detailing Tips, Schedules, Checklists, and Secrets Others Don't Tell You (Architectural Practice Simplified, 2nd edition)*
   http://www.ArchiteG.com

D. ***Planting Design Illustrated***
   http://outskirtspress.com/agent.php?key=11011&page=GangChen

ARE Mock Exam Series & ARE Exam Guide Series

ARE 5.0 Mock Exam Series

Practice Management (PcM) ARE 5.0 Mock Exam (Architect Registration Examination): *ARE 5.0 Overview, Exam Prep Tips, Hot Spots, Case Studies, Drag-and-Place, Solutions and Explanations.* **ISBN**: 9781612650388

Project Management (PjM) ARE 5.0 Mock Exam (Architect Registration Examination): *ARE 5.0 Overview, Exam Prep Tips, Hot Spots, Case Studies, Drag-and-Place, Solutions and Explanations.* **ISBN**: 9781612650371

Programming & Analysis (PA) ARE 5.0 Mock Exam (Architect Registration Exam): *ARE 5.0 Overview, Exam Prep Tips, Hot Spots, Case Studies, Drag-and-Place, Solutions and Explanations.* **ISBN**: 9781612650326

Project Planning & Design (PPD) ARE 5.0 Mock Exam (Architect Registration Examination): *ARE 5.0 Overview, Exam Prep Tips, Hot Spots, Case Studies, Drag-and-Place, Solutions and Explanations.* **ISBN**: 9781612650296

Project Development & Documentation (PDD) ARE 5.0 Mock Exam (Architect Registration Examination): *ARE 5.0 Overview, Exam Prep Tips, Hot Spots, Case Studies, Drag-and-Place, Solutions and Explanations*
ISBN: 9781612650258

Construction & Evaluation (CE) ARE 5.0 Mock Exam (Architect Registration Examination): *ARE 5.0 Overview, Exam Prep Tips, Hot Spots, Case Studies, Drag-and-Place, Solutions and Explanations*
ISBN: 9781612650241

Mock California Supplemental Exam (CSE of Architect Registration Examination): *CSE Overview, Exam Prep Tips, General Section and Project Scenario Section, Questions, Solutions and Explanations.* **ISBN**: 9781612650159

ARE 5.0 Exam Guide Series

Practice Management (PcM) ARE 5.0 Exam Guide (Architect Registration Examination): *ARE 5.0 Overview, Exam Prep Tips, Guide, and Critical Content.* **ISBN**: 9781612650333

Other books in the ARE 5.0 Exam Guide Series are being produced. Our goal is to produce one mock exam book *plus* one guidebook for each of the ARE 5.0 exam divisions. See the following link for the latest information:
http://www.GreenExamEducation.com

LEED Exam Guides series: Comprehensive Study Materials, Sample Questions, Mock Exam, Building LEED Certification and Going Green

LEED (Leadership in Energy and Environmental Design) is the most important trend of development, and it is revolutionizing the construction industry. It has gained tremendous momentum and has a profound impact on our environment.

From LEED Exam Guides series, you will learn how to

1. Pass the LEED Green Associate Exam and various LEED AP + exams (each book will help you with a specific LEED exam).

2. Register and certify a building for LEED certification.

3. Understand the intent for each LEED prerequisite and credit.

4. Calculate points for a LEED credit.

5. Identify the responsible party for each prerequisite and credit.

6. Earn extra credit (exemplary performance) for LEED.

7. Implement the local codes and building standards for prerequisites and credit.

8. Receive points for categories not yet clearly defined by USGBC.

There is currently NO official book on the LEED Green Associate Exam, and most of the existing books on LEED and LEED AP are too expensive and too complicated to be practical and helpful. The pocket guides in LEED Exam Guides series fill in the blanks, demystify LEED, and uncover the tips, codes, and jargon for LEED as well as the true meaning of "going green." They will set up a solid foundation and fundamental framework of LEED for you. Each book in the LEED Exam Guides series covers every aspect of one or more specific LEED rating system(s) in plain and concise language and makes this information understandable to all people.

These pocket guides are small and easy to carry around. You can read them whenever you have a few extra minutes. They are indispensable books for all people—administrators; developers; contractors; architects; landscape architects; civil, mechanical, electrical, and plumbing engineers; interns; drafters; designers; and other design professionals.

Why is the LEED Exam Guides series needed?

A number of books are available that you can use to prepare for the LEED exams:

1. *USGBC Reference Guides*. You need to select the correct version of the *Reference Guide* for your exam.

   The *USGBC Reference Guides* are comprehensive, but they give too much information. For example, *The LEED 2009 Reference Guide for Green Building Design and Construction (BD&C)* has about 700 oversized pages. Many of the calculations in the books are too detailed for the exam. They are also expensive (approximately $200 each, so most people may not buy them for their personal use, but instead, will seek to share an office copy).

   It is good to read a reference guide from cover to cover if you have the time. The problem is not too many people have time to read the whole reference guide. Even if you do read the whole guide, you may not remember the important issues to pass the LEED exam. You need to reread the material several times before you can remember much of it.

   Reading the reference guide from cover to cover without a guidebook is a difficult and inefficient way of preparing for the LEED AP Exam, because you do NOT know what USGBC and GBCI are looking for in the exam.

2. The USGBC workshops and related handouts are concise, but they do not cover extra credits (exemplary performance). The workshops are expensive, costing approximately $450 each.

3. Various books published by a third party are available on Amazon, bn.com and books.google.com. However, most of them are not very helpful.

   There are many books on LEED, but not all are useful.

   LEED Exam Guides series will fill in the blanks and become a valuable, reliable source:

   a. They will give you more information for your money. Each of the books in the LEED Exam Guides series has more information than the related USGBC workshops.

   b. They are exam-oriented and more effective than the USGBC reference guides.

   c. They are better than most, if not all, of the other third-party books. They give you comprehensive study materials, sample questions and answers, mock exams and answers, and critical information on building LEED certification and going green. Other third-party books only give you a fraction of the information.

   d. They are comprehensive yet concise. They are small and easy to carry around. You can read them whenever you have a few extra minutes.

   e. They are great timesavers. I have highlighted the important information that you need to understand and MEMORIZE. I also make some acronyms and short sentences to help you easily remember the credit names.

It should take you about 1 or 2 weeks of full-time study to pass each of the LEED exams. I have met people who have spent 40 hours to study and passed the exams.

You can find sample texts and other information on the LEED Exam Guides series in customer discussion sections under each of my book's listing on Amazon, bn.com and books.google.com.

What others are saying about *LEED GA Exam Guide* (Book 2, LEED Exam Guide series):

"Finally! A comprehensive study tool for LEED GA Prep!

"I took the 1-day Green LEED GA course and walked away with a power point binder printed in very small print—which was missing MUCH of the required information (although I didn't know it at the time). I studied my little heart out and took the test, only to fail it by 1 point. Turns out I did NOT study all the material I needed to in order to pass the test. I found this book, read it, marked it up, retook the test, and passed it with a 95%. Look, we all know the LEED GA exam is new and the resources for study are VERY limited. This one is the VERY best out there right now. I highly recommend it."
—ConsultantVA

"Complete overview for the LEED GA exam

"I studied this book for about 3 days and passed the exam ... if you are truly interested in learning about the LEED system and green building design, this is a great place to start."
—K.A. Evans

"A Wonderful Guide for the LEED GA Exam

"After deciding to take the LEED Green Associate exam, I started to look for the best possible study materials and resources. From what I thought would be a relatively easy task, it turned into a tedious endeavor. I realized that there are vast amounts of third-party guides and handbooks. Since the official sites offer little to no help, it became clear to me that my best chance to succeed and pass this exam would be to find the most comprehensive study guide that would not only teach me the topics, but would also give me a great background and understanding of what LEED actually is. Once I stumbled upon Mr. Chen's book, all my needs were answered. This is a great study guide that will give the reader the most complete view of the LEED exam and all that it entails.

"The book is written in an easy-to-understand language and brings up great examples, tying the material to the real world. The information is presented in a coherent and logical way, which optimizes the learning process and does not go into details that will not be needed for the LEED Green Associate Exam, as many other guides do. This book stays dead on topic and keeps the reader interested in the material.

"I highly recommend this book to anyone that is considering the LEED Green Associate Exam. I learned a great deal from this guide, and I am feeling very confident about my chances for passing my upcoming exam."
—Pavel Geystrin

"Easy to read, easy to understand

"I have read through the book once and found it to be the perfect study guide for me. The author does a great job of helping you get into the right frame of mind for the content of the exam. I had started by studying the Green Building Design and Construction reference guide for LEED projects produced by the USGBC. That was the wrong approach, simply too much information with very little retention. At 636 pages in textbook format, it would have been a daunting task to get through it. Gang Chen breaks down the points, helping to minimize the amount of information but maximizing the content I was able to absorb. I plan on going through the book a few more times, and I now believe I have the right information to pass the LEED Green Associate Exam."
—Brian Hochstein

"All in one—LEED GA prep material

"Since the LEED Green Associate exam is a newer addition by USGBC, there is not much information regarding study material for this exam. When I started looking around for material, I got really confused about what material I should buy. This LEED GA guide by Gang Chen is an answer to all my worries! It is a very precise book with lots of information, like how to approach the exam, what to study and what to skip, links to online material, and tips and tricks for passing the exam. It is like the 'one stop shop' for the LEED Green Associate Exam. I think this book can also be a good reference guide for green building professionals. A must-have!"
—SwatiD

"An ESSENTIAL LEED GA Exam Reference Guide

"This book is an invaluable tool in preparation for the LEED Green Associate (GA) Exam. As a practicing professional in the consulting realm, I found this book to be all-inclusive of the preparatory material needed for sitting the exam. The information provides clarity to the fundamental and advanced concepts of what LEED aims to achieve. A tremendous benefit is the connectivity of the concepts with real-world applications.

"The author, Gang Chen, provides a vast amount of knowledge in a very clear, concise, and logical media. For those that have not picked up a textbook in a while, it is very manageable to extract the needed information from this book. If you are taking the exam, do yourself a favor and purchase a copy of this great guide. Applicable fields: Civil Engineering, Architectural Design, MEP, and General Land Development."
—Edwin L. Tamang

Note: Other books in the **LEED Exam Guides series** are in the process of being produced. At least **one book will eventually be produced for each of the LEED exams.** The series include:

LEED v4 Green Associate Exam Guide (LEED GA): *Comprehensive Study Materials, Sample Questions, Mock Exam, Green Building LEED Certification, and Sustainability*, LEED Exam Guide series, ArchiteG.com. Latest Edition.

LEED GA MOCK EXAMS (LEED v4): *Questions, Answers, and Explanations: A Must-Have for the LEED Green Associate Exam, Green Building LEED Certification, and Sustainability,* LEED Exam Guide series, ArchiteG.com. Latest Edition

LEED v4 BD&C EXAM GUIDE: *A Must-Have for the LEED AP BD+C Exam: Comprehensive Study Materials, Sample Questions, Mock Exam, Green Building Design and Construction, LEED Certification, and Sustainability*, LEED Exam Guide series, ArchiteG.com. Latest Edition.

LEED v4 BD&C MOCK EXAMS: *Questions, Answers, and Explanations: A Must-Have for the LEED AP BD+C Exam, Green Building LEED Certification, and Sustainability*, LEED Exam Guide series, ArchiteG.com. Latest Edition.

LEED ID&C Exam Guide: *A Must-Have for the LEED AP ID+C Exam: Study Materials, Sample Questions, Green Interior Design and Construction, Green Building LEED Certification, and Sustainability,* LEED Exam Guide series, ArchiteG.com. Latest Edition.

LEED ID&C Mock Exam: *Questions, Answers, and Explanations: A Must-Have for the LEED AP ID+C Exam, Green Interior Design and Construction, Green Building LEED Certification, and Sustainability,* LEED Exam Guide series, ArchiteG.com. Latest Edition.

LEED O&M MOCK EXAMS: *Questions, Answers, and Explanations: A Must-Have for the LEED O&M Exam, Green Building LEED Certification, and Sustainability*, LEED Exam Guide series, ArchiteG.com. Latest Edition.

LEED O&M EXAM GUIDE: *A Must-Have for the LEED AP O+M Exam: Comprehensive Study Materials, Sample Questions, Mock Exam, Green Building Operations and Maintenance, LEED Certification, and Sustainability*, LEED Exam Guide series, ArchiteG.com. Latest Edition.

LEED HOMES EXAM GUIDE: *A Must-Have for the LEED AP Homes Exam: Comprehensive Study Materials, Sample Questions, Mock Exam, Green Building LEED Certification, and Sustainability*, LEED Exam Guide series, ArchiteG.com. Latest Edition.

LEED ND EXAM GUIDE: *A Must-Have for the LEED AP Neighborhood Development Exam: Comprehensive Study Materials, Sample Questions, Mock Exam, Green Building LEED Certification, and Sustainability*, LEED Exam Guide series, ArchiteG.com. Latest Edition.

How to order these books:
You can order the books listed above at:
http://www.GreenExamEducation.com

OR
http://www.ArchiteG.com

Building Construction

Project Management, Construction Administration, Drawings, Specs, Detailing Tips, Schedules, Checklists, and Secrets Others Don't Tell You (Architectural Practice Simplified, 2nd edition)

Learn the Tips, Become One of Those Who Know Building Construction and Architectural Practice, and Thrive!

For architectural practice and building design and construction industry, there are two kinds of people: those who know, and those who don't. The tips of building design and construction and project management have been undercover—until now.

Most of the existing books on building construction and architectural practice are too expensive, too complicated, and too long to be practical and helpful. This book simplifies the process to make it easier to understand and uncovers the tips of building design and construction and project management. It sets up a solid foundation and fundamental framework for this field. It covers every aspect of building construction and architectural practice in plain and concise language and introduces it to all people. Through practical case studies, it demonstrates the efficient and proper ways to handle various issues and problems in architectural practice and building design and construction industry.

It is for ordinary people and aspiring young architects as well as seasoned professionals in the construction industry. For ordinary people, it uncovers the tips of building construction; for aspiring architects, it works as a construction industry survival guide and a guidebook to shorten the process in mastering architectural practice and climbing up the professional ladder; for seasoned architects, it has many checklists to refresh their memory. It is an indispensable reference book for ordinary people, architectural students, interns, drafters, designers, seasoned architects, engineers, construction administrators, superintendents, construction managers, contractors, and developers.

You will learn:

1. How to develop your business and work with your client.
2. The entire process of building design and construction, including programming, entitlement, schematic design, design development, construction documents, bidding, and construction administration.
3. How to coordinate with governing agencies, including a county's health department and a city's planning, building, fire, public works departments, etc.
4. How to coordinate with your consultants, including soils, civil, structural, electrical, mechanical, plumbing engineers, landscape architects, etc.
5. How to create and use your own checklists to do quality control of your construction documents.
6. How to use various logs (i.e., RFI log, submittal log, field visit log, etc.) and lists (contact list, document control list, distribution list, etc.) to organize and simplify your work.
7. How to respond to RFI, issue CCDs, review change orders, submittals, etc.
8. How to make your architectural practice a profitable and successful business.

Planting Design Illustrated

A Must-Have for Landscape Architecture: A Holistic Garden Design Guide with Architectural and Horticultural Insight, and Ideas from Famous Gardens in Major Civilizations

One of the most significant books on landscaping!

This is one of the most comprehensive books on planting design. It fills in the blanks of the field and introduces poetry, painting, and symbolism into planting design. It covers in detail the two major systems of planting design: formal planting design and naturalistic planting design. It has numerous line drawings and photos to illustrate the planting design concepts and principles. Through in-depth discussions of historical precedents and practical case studies, it uncovers the fundamental design principles and concepts, as well as the underpinning philosophy for planting design. It is an indispensable reference book for landscape architecture students, designers, architects, urban planners, and ordinary garden lovers.

What Others Are Saying about *Planting Design Illustrated* ...

"I found this book to be absolutely fascinating. You will need to concentrate while reading it, but the effort will be well worth your time."
—Bobbie Schwartz, former president of APLD (Association of Professional Landscape Designers) and author of *The Design Puzzle: Putting the Pieces Together*.

"This is a book that you have to read, and it is more than well worth your time. Gang Chen takes you well beyond what you will learn in other books about basic principles like color, texture, and mass."
—Jane Berger, editor & publisher of gardendesignonline

"As a longtime consumer of gardening books, I am impressed with Gang Chen's inclusion of new information on planting design theory for Chinese and Japanese gardens. Many gardening books discuss the beauty of Japanese gardens, and a few discuss the unique charms of Chinese gardens, but this one explains how Japanese and Chinese history, as well as geography and artistic traditions, bear on the development of each country's style. The material on traditional Western garden planting is thorough and inspiring, too. *Planting Design Illustrated* definitely rewards repeated reading and study. Any garden designer will read it with profit."
—Jan Whitner, editor of the *Washington Park Arboretum Bulletin*

"Enhanced with an annotated bibliography and informative appendices, *Planting Design Illustrated* offers an especially "reader friendly" and practical guide that makes it a very strongly recommended addition to personal, professional, academic, and community library gardening & landscaping reference collection and supplemental reading list."
—Midwest Book Review

"Where to start? *Planting Design Illustrated* is, above all, fascinating and refreshing! Not something the lay reader encounters every day, the book presents an unlikely topic in an easily digestible, easy-to-follow way. It is superbly organized with a comprehensive table of contents, bibliography, and appendices. The writing, though expertly informative, maintains its accessibility throughout and is a joy to read. The detailed and beautiful illustrations expanding on the concepts presented were my favorite portion. One of the finest books I've encountered in this contest in the past 5 years."

—Writer's Digest 16th Annual International Self-Published Book Awards Judge's Commentary

"The work in my view has incredible application to planting design generally and a system approach to what is a very difficult subject to teach, at least in my experience. Also featured is a very beautiful philosophy of garden design principles bordering poetry. It's my strong conviction that this work needs to see the light of day by being published for the use of professionals, students & garden enthusiasts."

—Donald C. Brinkerhoff, FASLA, chairman and CEO of Lifescapes International, Inc.

Index

Notes

Notes

Notes

Notes

www.ingramcontent.com/pod-product-compliance
Lightning Source LLC
LaVergne TN
LVHW082344200926
848481LV00012B/3447
* 9 7 8 1 6 1 2 6 5 0 1 5 9 *